Works Pencoyd Iron

Wrought iron and steel in construction. Convenient rules, formulae, and tables for the strength of wrought iron shapes used as beams, struts, shafts, etc.

Second Edition

Works Pencoyd Iron

Wrought iron and steel in construction. Convenient rules, formulae, and tables for the strength of wrought iron shapes used as beams, struts, shafts, etc.
Second Edition

ISBN/EAN: 9783337156664

Printed in Europe, USA, Canada, Australia, Japan

Cover: Foto ©berggeist007 / pixelio.de

More available books at **www.hansebooks.com**

WROUGHT IRON AND STEEL

IN

CONSTRUCTION.

CONVENIENT RULES, FORMULÆ, AND TABLES FOR THE
STRENGTH OF WROUGHT IRON SHAPES USED AS
BEAMS, STRUTS, SHAFTS, ETC., MANUFAC-
TURED BY THE PENCOYD IRON WORKS.

SECOND EDITION, REVISED AND ENLARGED.

NEW YORK.
JOHN WILEY & SONS,
15 ASTOR PLACE,
1885.

PRESS OF J. J. LITTLE & CO.,
NOS. 10 TO 20 ASTOR PLACE, NEW YORK.

PREFACE.

To Engineers and Builders in Iron and Steel this volume is presented, with the hope that it may be of assistance to them in their daily labors, and afford information upon some points which have not heretofore been put in published form. It has been the aim of the author to eliminate as far as possible matters of theory from statements of facts, that, where conflict of opinion may arise, each one may draw his own conclusions. It was considered advisable to treat only of subjects relating to Iron and Steel, referring to any of the numerous engineers' pocket-books for information upon outside matters.

As far as possible, doubtful points were corroborated by experiments, and especially the article upon "Struts" is based upon the results of several hundred carefully conducted experiments at Pencoyd, for more detailed information concerning which we would refer to two papers by Mr. Jas. Christie, published in the Transactions of the American Society of Civil Engineers, entitled, "Experiments on the Strength of Wrought Iron Struts," and "The Strength and Elasticity of Structural Steel," wherein the above experiments are fully described. Hereafter should errors be detected by a more perfect knowledge of the physical properties of the materials treated of, we shall be glad to acknowledge the same, but now offer the following pages as the best results we are able to obtain from present practice.

A. & P. ROBERTS & CO.

PENCOYD, May, 1884.

PREFACE TO SECOND EDITION.

In preparing the Sécond Edition for the press we have corrected some small errors occurring in various places in the first edition, which were discovered after its publication. A few new tables of weights of separators for beams, of bolts, nuts and rivets, which were deemed useful in architectural calculations, have been added. Some additional shapes are described, and several old sections of beams and channels changed to more efficient forms, by better distribution of material in the flanges. At the present writing we have no alterations to make in our conclusions in regard to steel, our experiments up to date seeming to confirm our results as then announced.

A. & P. ROBERTS & CO.

Pencoyd, January, 1885.

PREFACE TO THIRD EDITION.

MORE than a year has elapsed since the publication of the first edition of this little volume, and we are now preparing a third for the press. A few new sections have been added and several errors overlooked in the earlier editions corrected, so that we believe very few, if any, now exist. Our conclusions in regard to struts, based upon Mr. Christie's experiments, have stood the test of publication and criticism, and we think at this day can be said to have more fully the stamp of authority than when first issued. We trust this Hand Book has and will continue to be of value to all who daily use wrought iron and steel in construction.

A. & P. ROBERTS & CO.

PENCOYD, July, 1885.

CONTENTS.

For a full detail of the contents see Index.

WROUGHT IRON AND STEEL IN CONSTRUCTION.

TABLES OF DIMENSIONS.

THE following tables give the principal dimensions of the standard shapes of structural iron and steel rolled at Pencoyd.

Further particulars of the sections will be found in the illustrations at the end of the book.

For beams and channels the least and greatest sections of each size are described in the preliminary tables. Any intermediate sectional areas between the maximum and minimum can be rolled, but the flanges remain unaltered, the web only being thickened. The weights per yard corresponding to increased web thicknesses are given in annexed tables. For angles, any thickness between the maximum and minimum can be rolled, corresponding weights for the principal intermediate thicknesses being given in the tables.

The legs of angles increase slightly in width as the thickness is increased. This renders the actual weights corresponding to given thickness somewhat uncertain. Therefore either the desired thickness or weight per yard should be specified, but not both. (The methods of altering the thickness of the foregoing sections, are illustrated in plate No. 28.) The cross-hatched sections represent the least areas, and the blank section the added thickness.

Tee sections cannot be altered from the standard as given in the tables. Flat bars can be rolled to any thickness between the limits given in the list.

SIZES OF MINIMUM AND MAXIMUM SECTIONS.

PENCOYD 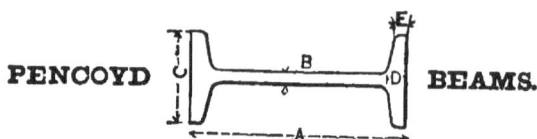 BEAMS.

Chart Number.	Size in Inches.	Minimum Weight per yard.	Maximum Weight per yard.	Minimum Web Thickness.	Maximum Web Thickness.	Minimum Flange Width.	Maximum Flange Width.	Flange Thickness.	Flange Thickness.
	A			B	B	C	C	D	E
1	15	200	233	$\frac{3}{8}$	$\frac{7}{8}$	$5\frac{3}{4}$	$5\frac{11}{32}$	$1\frac{5}{16}$	$\frac{3}{4}$
2	15	145	201	$\frac{7}{16}$	$\frac{13}{16}$	$5\frac{5}{8}$	$5\frac{1}{4}$	$1\frac{1}{8}$	$\frac{5}{8}$
3	12	168	194	$\frac{3}{8}$	$\frac{7}{8}$	$5\frac{1}{2}$	$5\frac{19}{32}$	$1\frac{1}{4}$	$\frac{11}{16}$
4	12	120	163	$\frac{21}{64}$	$\frac{13}{16}$	$4\frac{51}{64}$	$5\frac{5}{32}$	1	$\frac{33}{64}$
5	$10\frac{1}{2}$	134	161	$\frac{15}{32}$	$\frac{23}{32}$	$5\frac{1}{4}$	$5\frac{1}{2}$	$1\frac{1}{8}$	$\frac{21}{32}$
5½	$10\frac{1}{2}$	108	135	$\frac{11}{32}$	$\frac{21}{32}$	$4\frac{7}{8}$	$5\frac{1}{8}$	$1\frac{5}{32}$	$\frac{17}{32}$
6	$10\frac{1}{2}$	89	109	$\frac{11}{32}$	$\frac{17}{32}$	$4\frac{1}{2}$	$4\frac{11}{16}$	$\frac{15}{32}$	$\frac{7}{16}$
7	10	112	137	$\frac{1}{2}$	$\frac{7}{8}$	$4\frac{5}{8}$	$4\frac{7}{8}$	$1\frac{1}{16}$	$\frac{1}{2}$
8	10	90	106	$\frac{11}{32}$	$\frac{5}{8}$	$4\frac{3}{8}$	$4\frac{17}{32}$	$\frac{21}{32}$	$\frac{15}{32}$
9	9	90	122	$\frac{13}{32}$	$\frac{3}{4}$	$4\frac{5}{8}$	$4\frac{23}{32}$	$\frac{21}{32}$	$\frac{1}{2}$
10	9	70	88	$\frac{19}{64}$	$\frac{1}{2}$	$4\frac{1}{8}$	$4\frac{21}{32}$	$\frac{15}{32}$	$\frac{5}{8}$
11	8	81	109	$\frac{13}{32}$	$\frac{3}{4}$	$4\frac{1}{4}$	$4\frac{15}{32}$	$\frac{27}{32}$	$\frac{15}{32}$
12	8	65	75	$\frac{5}{16}$	$\frac{7}{16}$	4	$4\frac{1}{8}$	$\frac{3}{4}$	$\frac{3}{8}$
13	7	65	88	$\frac{7}{16}$	$\frac{3}{4}$	$3\frac{13}{16}$	$4\frac{1}{8}$	$\frac{3}{4}$	$\frac{15}{32}$
14	7	51	88	$\frac{15}{64}$	$\frac{3}{4}$	$3\frac{13}{64}$	$4\frac{1}{8}$	$\frac{21}{32}$	$\frac{15}{32}$
15	6	50	63	$\frac{13}{32}$	$\frac{5}{8}$	$3\frac{5}{32}$	$3\frac{5}{8}$	$\frac{11}{16}$	$\frac{5}{16}$
16	6	40	63	$\frac{1}{4}$	$\frac{5}{8}$	3	$3\frac{1}{3}$	$\frac{21}{32}$	$\frac{5}{16}$
17	5	34	40	$\frac{5}{16}$	$\frac{7}{16}$	$2\frac{17}{32}$	$2\frac{17}{32}$	$\frac{1}{2}$	$\frac{1}{4}$
18	5	30	40	$\frac{7}{32}$	$\frac{7}{16}$	$2\frac{3}{8}$	$2\frac{17}{32}$	$\frac{1}{2}$	$\frac{1}{4}$
19	4	28	38	$\frac{1}{4}$	$\frac{1}{2}$	$2\frac{3}{4}$	3	$\frac{1}{2}$	$\frac{9}{32}$
20	4	18.5	21.5	$\frac{11}{64}$	$\frac{1}{4}$	$2\frac{1}{4}$	$2\frac{21}{64}$	$\frac{5}{8}$	$\frac{7}{32}$
21	3	23	28.6	$\frac{1}{4}$	$\frac{7}{16}$	$2\frac{1}{2}$	$2\frac{11}{16}$	$\frac{7}{16}$	$\frac{1}{4}$
22	3	17	21.7	$\frac{5}{32}$	$\frac{5}{16}$	$2\frac{1}{4}$	$2\frac{13}{32}$	$\frac{13}{32}$	$\frac{13}{64}$

The width of the flange varies directly with the thickness of the web.

WEIGHTS OF VARIOUS WEB THICKNESSES.

PENCOYD  **BEAMS.**

Chart Number.	Depth in Inches.	Minimum Web Thickness.	Minimum Weight per yard.	Approximate Weight in pounds per yard for each Thickness of Web in inches. (Calculated upon the basis that one Cubic Foot of Iron weighs 480 lbs.)							
				$\frac{1}{4}$	$\frac{5}{16}$	$\frac{3}{8}$	$\frac{7}{16}$	$\frac{1}{2}$	$\frac{5}{8}$	$\frac{3}{4}$	$\frac{7}{8}$
1	15	$\frac{3}{32}$	200.0	....	....	....	....	....	....	214.0	233.0
2	15	$\frac{7}{16}$	145.0	....	....	....	145.0	154.5	173.0	192.0	.....
3	12	$\frac{3}{32}$	168.0	....	....	....	....	....	....	179.0	194.0
4	12	$\frac{29}{64}$	120.0	....	....	....	....	125.5	140.5	155.5	.....
5	10½	$\frac{15}{32}$	134.4	....	....	....	.....	137.7	150.8	.....	.....
5½	10½	$\frac{13}{32}$	108.3	....	....	....	111.6	118.1	131.2	.....	.....
6	10½	$\frac{11}{32}$	89.3	....	....	92.6	99.2	105.8	.....	.....	.....
7	10	$\frac{1}{2}$	111.7	....	....	....	.....	111.7	124.2	136.7	.....
8	10	$\frac{11}{32}$	90.4	....	....	93.6	99.8	106.0	.....	.....	.....
9	9	$\frac{13}{32}$	90.7	....	....	....	93.5	99.1	110.4	121.6	.....
10	9	$\frac{15}{64}$	69.8	....	71.2	76.8	82.4	88.1	.....	.....	.....
11	8	$\frac{13}{32}$	81.4	....	....	....	83.9	88.9	98.9	108.9	.....
12	8	$\frac{5}{16}$	65.3	....	65.3	70.3	75.3	.....	.....	.....	.....
13	7	$\frac{7}{16}$	65.8	....	....	....	65.8	70.2	78.9	87.7	.....
14	7	$\frac{15}{64}$	51.4	52.5	56.9	61.2	65.6	70.0	78.7	87.5	.....
15	6	$\frac{13}{32}$	50.0	....	....	....	52.0	55.5	63.0	.....	.....
16	6	$\frac{1}{4}$	40.0	40.0	44.0	47.5	51.0	55.0	63.0	.....	.....
17	5	$\frac{5}{16}$	34.0	....	34.0	37.0	40.0	.....	.....	.....	.....
18	5	$\frac{7}{32}$	30.0	31.5	34.6	37.8	40.9	.....	.....	.....	.....
19	4	$\frac{1}{4}$	28.0	28.0	30.5	33.0	35.5	38.0	.....	.....	.....
20	4	$\frac{11}{64}$	18.5	21.5	....	....	.....	.....	.....	.....	.....
21	3	$\frac{1}{4}$	23.0	23.0	24.8	26.7	28.6	.....	.....	.....	.....
22	3	$\frac{1}{32}$	17.0	19.8	21.7	....	.....	.....	.....	.....	.....

Beams of any weight between the minimum and maximum weight per yard, given in the table, can be furnished.

SIZES OF MINIMUM AND MAXIMUM SECTIONS.

PENCOYD 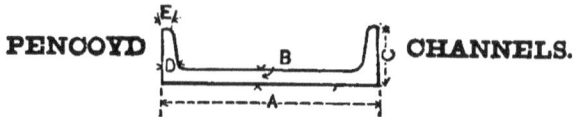 **CHANNELS.**

Chart Number.	Size in inches.	Minimum Weight per yard.	Maximum Weight per yard.	Minimum Web Thickness.	Maximum Web Thickness.	Minimum Flange Width.	Maximum Flange Width.	Flange Thickness.	Flange Thickness.
	A			B	B	C	C	D	E
30	15	139.0	204.5	$\frac{9}{16}$	1	4	$4\frac{3}{4}$	1	$\frac{5}{8}$
31	12	88.5	160.0	$\frac{13}{32}$	1	$2\frac{1}{2}$	$3\frac{1}{32}$	1	$\frac{9}{16}$
32	12	60.0	101.5	$\frac{9}{32}$	$\frac{5}{8}$	$2\frac{3}{8}$	$2\frac{5}{8}$	$\frac{3}{4}$	$\frac{13}{32}$
34	10	60.0	106.0	$\frac{3}{8}$	$\frac{3}{4}$	$2\frac{1}{2}$	$3\frac{1}{16}$	$\frac{11}{16}$	$\frac{7}{16}$
35	10	49.0	86.5	$\frac{1}{4}$	$\frac{5}{8}$	$2\frac{3}{8}$	$2\frac{3}{4}$	$\frac{3}{4}$	$\frac{3}{8}$
36	9	53.0	92.0	$\frac{5}{16}$	$\frac{3}{4}$	$2\frac{7}{16}$	$2\frac{7}{8}$	$\frac{3}{4}$	$\frac{13}{32}$
37	9	37.0	61.0	$\frac{15}{64}$	$\frac{1}{2}$	$2\frac{9}{64}$	$2\frac{13}{32}$	$\frac{35}{64}$	$\frac{13}{64}$
38	8	43.0	80.5	$\frac{9}{32}$	$\frac{3}{4}$	$2\frac{9}{32}$	$2\frac{3}{4}$	$\frac{43}{64}$	$\frac{11}{32}$
39	8	30.0	54.0	$\frac{13}{64}$	$\frac{1}{2}$	2	$2\frac{5}{64}$	$\frac{33}{64}$	$\frac{3}{8}$
40	7	41.0	73.0	$\frac{13}{64}$	$\frac{3}{4}$	$2\frac{19}{64}$	$2\frac{3}{4}$	$\frac{11}{16}$	$\frac{7}{16}$
41	7	26.0	49.0	$\frac{3}{16}$	$\frac{1}{2}$	$1\frac{13}{16}$	$2\frac{3}{16}$	$\frac{7}{16}$	$\frac{1}{4}$
42	6	31.9	54.4	$\frac{1}{4}$	$\frac{5}{8}$	$2\frac{1}{4}$	$2\frac{5}{8}$	$\frac{9}{16}$	$\frac{3}{8}$
43	6	27.6	50.1	$\frac{1}{4}$	$\frac{5}{8}$	2	$2\frac{3}{8}$	$\frac{33}{64}$	$\frac{1}{4}$
44	6	22.7	39.6	$\frac{7}{32}$	$\frac{1}{2}$	$1\frac{3}{4}$	$2\frac{3}{32}$	$\frac{3}{8}$	$\frac{5}{16}$
45	5	27.3	46.0	$\frac{1}{4}$	$\frac{5}{8}$	2	$2\frac{3}{8}$	$\frac{19}{32}$	$\frac{3}{8}$
46	5	18.8	32.9	$\frac{7}{32}$	$\frac{1}{2}$	$1\frac{5}{8}$	$1\frac{33}{32}$	$\frac{3}{8}$	$\frac{5}{16}$
47	4	21.5	31.5	$\frac{1}{4}$	$\frac{1}{2}$	$1\frac{13}{32}$	$1\frac{11}{32}$	$\frac{17}{32}$	$\frac{1}{4}$
48	4	17.5	23.7	$\frac{7}{32}$	$\frac{3}{8}$	$1\frac{3}{16}$	$1\frac{11}{32}$	$\frac{11}{32}$	$\frac{1}{4}$
49	3	15.2	18.9	$\frac{7}{32}$	$\frac{11}{32}$	$1\frac{1}{32}$	$1\frac{11}{32}$	$\frac{11}{32}$	$\frac{1}{4}$
50	$2\frac{1}{2}$	11.3	11.3	$\frac{1}{4}$	$\frac{1}{4}$	$1\frac{3}{8}$	$1\frac{3}{8}$	$\frac{1}{4}$	$\frac{1}{4}$
51	2	8.75	10.0	$\frac{7}{32}$	$\frac{9}{32}$	$1\frac{3}{32}$	$1\frac{5}{32}$	$\frac{5}{16}$	$\frac{3}{16}$
52	$1\frac{1}{4}$	3.5	3.5	$\frac{3}{32}$	$\frac{3}{32}$	$1\frac{9}{32}$	$1\frac{9}{32}$	$\frac{1}{4}$	$\frac{3}{32}$

The width of the flange varies directly with the thickness of the web.

WEIGHTS OF VARIOUS WEB THICKNESSES.

PENCOYD CHANNELS.

Chart Number.	Depth in inches.	Minimum Web Thickness.	Minimum Weight per yard.	Approximate Weight in Pounds per Yard for each Thickness of Web in inches. Calculated upon the basis that one Cubic Foot of Iron weighs 480 lbs.								
				$\frac{1}{4}$	$\frac{5}{16}$	$\frac{3}{8}$	$\frac{7}{16}$	$\frac{1}{2}$	$\frac{5}{8}$	$\frac{3}{4}$	$\frac{7}{8}$	1
30	15	$\frac{9}{16}$	139.0	....	....	....	....	.....	148.0	167.0	186.0	204.5
31	12	$\frac{13}{32}$	88.5	....	....	....	92.0	100.0	115.0	130.0	145.0	160.0
32	12	$\frac{3}{32}$	60.0	....	64.0	71.5	79.0	86.5	101.5	....	....	....
34	10	$\frac{9}{32}$	60.0	....	63.0	69.0	75.5	81.0	94.0	106.0	....	....
35	10	$\frac{1}{4}$	49.0	49.0	55.0	61.5	67.0	74.0	86.5	....	....	....
36	9	$\frac{5}{16}$	53.0	....	53.0	58.5	64.0	70.0	81.0	92.0	....	....
37	9	$\frac{15}{64}$	37.0	....	44.0	49.5	55.0	61.0	....	....		
38	8	$\frac{9}{32}$	43.0	....	45.5	50.5	55.5	60.5	70.5	80.5	....	....
39	8	$\frac{11}{64}$	30.0	34.0	39.0	44.0	49.0	54.0	....	....		
40	7	$\frac{19}{64}$	41.0	....	42.0	46.5	51.0	55.0	64.0	73.0		
41	7	$\frac{3}{16}$	26.0	31.5	36.0	40.0	44.5	49.0	....			
42	6	$\frac{1}{4}$	31.9	31.9	35.6	39.4	43.1	46.9	54.4	....	....	....
43	6	$\frac{1}{4}$	27.6	27.6	31.4	35.1	38.9	42.6	50.1	....	....	....
44	6	$\frac{7}{32}$	22.7	24.6	28.3	32.1	35.8	39.6	....	....	....	....
45	5	$\frac{1}{4}$	27.3	27.3	30.4	33.5	36.7	39.8	46.0	....	....	....
46	5	$\frac{7}{32}$	18.8	20.4	23.5	26.6	29.7	32.9	....	....	....	....
47	4	$\frac{1}{4}$	21.5	21.5	24.0	26.5	29.0	31.5	....			
48	4	$\frac{7}{32}$	17.5	18.7	21.2	23.7	....	....	....			
49	3	$\frac{7}{32}$	15.2	16.1	18.0	....	....	....				
50	2¼	$\frac{1}{4}$	11.3	11.3	....	....	....	....				
51	2	$\frac{7}{32}$	8.75	9.4	....	....	....					
52	1¾	$\frac{3}{32}$	3.5	....	....	....						

Channels of any weight between the minimum and maximum weight per yard, given in the table, can be furnished.

SIZES OF MINIMUM AND MAXIMUM SECTIONS.

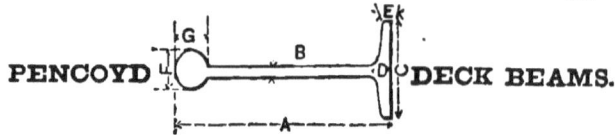

PENCOYD   DECK BEAMS.

Chart Number.	Size in inches.	Minimum Weight per yard.	Maximum Weight per yard.	Minimum Web Thickness.	Maximum Web Thickness.	Minimum Flange Width.	Maximum Flange Width.	Minimum Bulb Width.	Maximum Bulb Width.	Bulb Depth.	Flange Thickness.	Flange Thickness.
	A			B	B	C	C	F	F	G	D	E
60	12	104.0	138.0	$\frac{13}{32}$	$1\frac{1}{16}$	$5\frac{3}{4}$	$6\frac{1}{32}$	$2\frac{1}{8}$	$2\frac{13}{32}$	$1\frac{5}{8}$	$\frac{25}{32}$	$\frac{15}{32}$
61	11	91.0	118.0	$\frac{3}{8}$	$\frac{5}{8}$	$5\frac{1}{2}$	$5\frac{3}{4}$	2	$2\frac{1}{4}$	$1\frac{1}{2}$	$\frac{3}{4}$	$\frac{7}{16}$
62	10	80.0	105.0	$\frac{3}{8}$	$\frac{5}{8}$	$5\frac{1}{4}$	$5\frac{1}{2}$	$1\frac{7}{8}$	$2\frac{1}{8}$	$1\frac{13}{32}$	$\frac{11}{16}$	$\frac{13}{32}$
63	9	72.0	94.0	$\frac{3}{8}$	$\frac{5}{8}$	5	$5\frac{1}{4}$	$1\frac{25}{32}$	$2\frac{3}{32}$	$1\frac{13}{32}$	$\frac{5}{8}$	$\frac{3}{8}$
64	8	61.0	84.0	$\frac{11}{32}$	$\frac{5}{8}$	$4\frac{5}{8}$	$4\frac{29}{32}$	$1\frac{11}{16}$	$1\frac{31}{32}$	$1\frac{5}{16}$	$\frac{19}{32}$	$\frac{11}{32}$
65	7	52.0	72.0	$\frac{11}{32}$	$\frac{5}{8}$	$4\frac{1}{4}$	$4\frac{17}{32}$	$1\frac{9}{16}$	$1\frac{27}{32}$	$1\frac{3}{16}$	$\frac{9}{16}$	$\frac{5}{16}$
66	6	42.0	57.0	$\frac{5}{16}$	$\frac{9}{16}$	$3\frac{3}{4}$	4	$1\frac{7}{16}$	$1\frac{11}{16}$	$1\frac{1}{16}$	$\frac{17}{32}$	$\frac{9}{32}$
67	5	34.0	46.0	$\frac{5}{16}$	$\frac{9}{16}$	$3\frac{1}{4}$	$3\frac{1}{2}$	$1\frac{5}{16}$	$1\frac{9}{16}$	$1\frac{5}{16}$	$\frac{1}{2}$	$\frac{1}{4}$

WEIGHTS OF VARIOUS WEB THICKNESSES.

PENCOYD 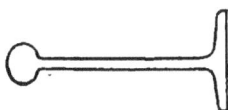 **DECK BEAMS.**

Depth in inches.	Minimum Web Thickness.	Minimum Weight per yard.	Approximate Weight in pounds per yard for each Thickness of Web in inches. Calculated upon the basis that one Cubic Foot of Iron weighs 480 lbs.							
			$\frac{1}{4}$	$\frac{5}{16}$	$\frac{3}{8}$	$\frac{7}{16}$	$\frac{1}{2}$	$\frac{9}{16}$	$\frac{5}{8}$	$\frac{11}{16}$
12	$\frac{1}{2}$	104.0	....	....	....	108.0	115.0	123.0	130.0	138.0
11	$\frac{3}{8}$	91.0	....	....	91.0	98.0	105.0	111.0	118.0	......
10	$\frac{3}{8}$	80.0	....	....	80.0	86.0	92.0	99.0	105.0	......
9	$\frac{3}{8}$	72.0	....	....	72.0	77.0	83.0	89.0	94.0	......
8	$\frac{11}{32}$	61.0	....	....	64.0	69.0	74.0	79.0	84.0	......
7	$\frac{11}{32}$	52.0	....	....	54.0	58.0	63.0	67.0	72.0	......
6	$\frac{5}{16}$	42.0	....	42.0	46.0	49.0	53.0	57.0	......	......
5	$\frac{5}{16}$	34.0	....	34.0	37.0	40.0	43.0	46.0	......	......

PENCOYD ANGLES.

EVEN LEGS.

WEIGHTS PER YARD OF VARIOUS THICKNESSES.

One cubic foot weighing 480 lbs.

Chart Number.	Size in Inches.	$\frac{3}{8}''$	$\frac{7}{16}''$	$\frac{1}{2}''$	$\frac{9}{16}''$	$\frac{5}{8}''$	$1\frac{1}{16}''$	$\frac{3}{4}''$	$1\frac{3}{16}''$	$\frac{7}{8}''$	$1''$
120	6×6	...	50.6	57.5	64.3	71.1	77.8	84.4	90.6	97.3	110.0
121	5×5	....	41.8	47.5	53.1	58.6	64.0	69.4	74.7	79.8	90.0
122	4×4	28.6	33.1	37.5	41.8	46.1	50.3	54.4	....	....	......
123	$3\frac{1}{2} \times 3\frac{1}{2}$	24.8	28.7	32.5	36.2	39.8	....	....	....	....	......

Chart Number.	Size in Inches.	$\frac{1}{8}''$	$\frac{3}{16}''$	$\frac{1}{4}''$	$\frac{5}{16}''$	$\frac{3}{8}''$	$\frac{7}{16}''$	$\frac{1}{2}''$	$\frac{9}{16}''$	$\frac{5}{8}''$	
124	3×3	....	...	14.4	17.8	21.1	24.3	27.5	30.6	33.6	......
125	$2\frac{3}{4} \times 2\frac{3}{4}$	....	....	13.1	16.2	19.2	22.1	25.0	....	....	......
126	$2\frac{1}{2} \times 2\frac{1}{2}$	....	....	11.9	14.6	17.3	19.9	22.5	....	....	......
127	$2\frac{1}{4} \times 2\frac{1}{4}$	....	....	10.6	13.1	15.5	17.8	....	....	....	......
128	2×2	....	7.1	9.4	11.5	13.6	....	....	....	....	......
129	$1\frac{3}{4} \times 1\frac{3}{4}$	....	6.2	8.1	9.9	11.7	....	....	....	....	......
130	$1\frac{1}{2} \times 1\frac{1}{2}$	....	5.3	6.9	8.4	9.8	....	....	....	....	......
131	$1\frac{1}{4} \times 1\frac{1}{4}$	3.0	4.3	5.6	....	....	....	....	....	....	......
132	1×1	2.3	3.4	4.4	....	....	....	....	....	....	......

PENCOYD ANGLES.

UNEVEN LEGS.

WEIGHTS PER YARD OF VARIOUS THICKNESSES.

One cubic foot weighing 480 lbs.

Chart Number.	Size in Inches.	$\frac{1}{4}$	$\frac{5}{16}$	$\frac{3}{8}$	$\frac{7}{16}$	$\frac{1}{2}$	$\frac{9}{16}$	$\frac{5}{8}$	$\frac{3}{4}$	$\frac{7}{8}$	1
140	6 ×4	....	....	....	41.8	47.5	53.0	58.6	69.4	79.8	90.0
141	5 ×4	....	....	32.3	37.4	42.5	47.4	52.3	61.8	71.1	80.0
142	5 ×3½	....	....	30.5	35.1	40.0	44.6	49.2	58.1	....	....
143	5 ×3	....	....	28.6	33.0	37.5	41.7	46.0	54.4	....	....
144	4½×3	....	....	26.7	30.9	35.0	39.0	43.0	....	....	....
145	4 ×3½	....	....	26.7	30 9	35.0	39.0	43.0	....	....	
146	4 ×3	....	21.0	24.8	28.7	32.5	36.2	39.8	....	....	....
147	3½×3	....	....	23.0	26.5	30.0	33.4	36.7	....	....	....
148	3 ×2½	....	16.2	19.2	22.1	25.0	....		....	....	
149	3 ×2	11.9	14.6	17.3	19.9	22.5	...				
150	3½×2½	....	17.8	21.1	24 3	27.5		...	....		
151	6 ×3½	....	....	34.5	39.6	45.0	50.3	55.5	65.6	75.5	85.0
152	6½×4	....	....	....	44.0	50.0	55.9	61.7	73.1	84.2	95.0
153	5½×3½	....	....	32.3	37.4	42.5	47.4	52.3	...	....	....
154	7 ×3½	....	....	...	....	....	...	61.7	73.1	84.2	95.0
155	2½×2	10.6	13.1	15.4	17.7	20.0	....	....	....		
156	2¼×1½	8.7	10.7	12.6	....	....	....		....		
157	2 ×1¼	7.5	9.2	10.8	....	....		....	....		

PENCOYD ANGLES.

SQUARE ROOT.

WEIGHTS PER YARD OF VARIOUS THICKNESSES.

One cubic foot weighing 480 lbs.

CHART NUMBER.	SIZE IN INCHES.	$\frac{1}{8}$	$\frac{3}{16}$	$\frac{1}{4}$	$\frac{5}{16}$	$\frac{3}{8}$	$\frac{7}{16}$	$\frac{1}{2}$	$\frac{9}{16}$	$\frac{5}{8}$
160	4×4	....	....	....	....	28.6	33.0	37.6	41.8	46.0
161	$3\frac{1}{2} \times 3\frac{1}{2}$	....	....	....	20.8	24.8	28.7	32.5	....	....
162	3×3	....	....	14.4	17.8	21.2	24.4	27.5	....	....
163	$2\frac{3}{4} \times 2\frac{3}{4}$	....	....	13.1	16.2	19.2	22.1	25.0	....	....
164	$2\frac{1}{2} \times 2\frac{1}{2}$	....	....	11.9	14.6	17.3	19.9	....	....	....
165	$2\frac{1}{4} \times 2\frac{1}{4}$	....	....	10.6	13.1	15.5	17.8	....	....	....
166	2×2	....	....	9.4	11.5	13.6	....	....	...	....
167	$1\frac{3}{4} \times 1\frac{3}{4}$	....	...	8.1	9.9	11.7	....	....	....	...
168	$1\frac{1}{2} \times 1\frac{1}{2}$	....	5.3	6.9	8.4	....	....	....		....
169	$1\frac{1}{4} \times 1\frac{1}{4}$	....	4.3	5.6	7.0	....	....	....	....	....
170	1×1	2.3	3.4	4.4	....	....	....	....	...	....
171	$1\frac{1}{2} \times \frac{15}{16}$	....	....	5.9	....	....	....	....	....	....

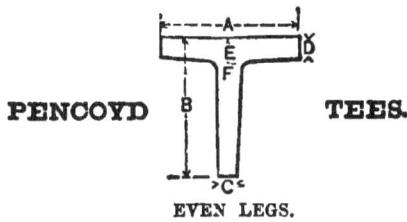

PENCOYD TEES.

EVEN LEGS.

Chart Number.	Width of base.	Height of stem.	Thickness of base.	Thickness of base.	Thickness of stem.	Thickness of stem.	WEIGHT PER YARD.
	A	B	D	E	C	F	
70	4	4	$\frac{7}{16}$	$\frac{9}{16}$	$\frac{7}{16}$	$\frac{1}{2}$	36.5 lbs.
71	3½	3½	$\frac{7}{16}$	$\frac{1}{2}$	$\frac{7}{16}$	$\frac{1}{2}$	31.0 lbs.
72	3	3	$\frac{13}{32}$	$\frac{15}{32}$	$\frac{7}{16}$	$\frac{1}{2}$	26.0 lbs.
73	2½	2½	$\frac{13}{32}$	$\frac{15}{32}$	$\frac{13}{32}$	$\frac{15}{32}$	19.5 lbs.
74	2½	2½	$\frac{11}{32}$	$\frac{13}{32}$	$\frac{11}{32}$	$\frac{13}{32}$	17.52 lbs.
75	2¼	2¼	$\frac{1}{4}$	$\frac{3}{8}$	$\frac{1}{4}$	$\frac{1}{4}$	11.75 lbs.
76	2¼	2¼	$\frac{1}{4}$	$\frac{5}{16}$	$\frac{1}{4}$	$\frac{5}{16}$	12.0 lbs.
77	2	2	$\frac{1}{4}$	$\frac{5}{16}$	$\frac{1}{4}$	$\frac{5}{16}$	10.5 lbs.
78	1¾	1¾	$\frac{3}{16}$	$\frac{1}{4}$	$\frac{3}{16}$	$\frac{1}{4}$	7.1 lbs.
79	1½	1½	$\frac{3}{16}$	$\frac{1}{4}$	$\frac{3}{16}$	$\frac{1}{4}$	6.0 lbs.
80	1¼	1¼	$\frac{3}{16}$	$\frac{1}{4}$	$\frac{3}{16}$	$\frac{1}{4}$	4.5 lbs.
81	1	1	$\frac{5}{32}$	$\frac{1}{4}$	$\frac{5}{32}$	$\frac{1}{4}$	3.0 lbs.
82	3	3	$\frac{5}{16}$	$\frac{3}{8}$	$\frac{5}{16}$	$\frac{3}{8}$	19.3 lbs.
83	3	3	$\frac{3}{8}$	$\frac{7}{16}$	$\frac{3}{8}$	$\frac{7}{16}$	22.6 lbs.

Weights of these sections cannot be varied.

PENCOYD 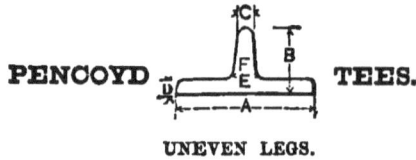 TEES.

UNEVEN LEGS.

Chart Number.	Width of base.	Height of stem.	Thickness of base.	Thickness of base.	Thickness of stem.	Thickness of stem.	WEIGHT PER YARD.
	A	B	D	E	C	F	
90	$4\frac{1}{2}$	$3\frac{1}{2}$	$\frac{7}{16}$	$\frac{1}{2}$	$1\frac{1}{6}$	$\frac{7}{8}$	44.5 lbs.
91	4	$3\frac{1}{2}$	$\frac{7}{16}$	$\frac{3}{4}$	$\frac{4}{16}$	$\frac{3}{4}$	41.8 lbs.
92	5	$2\frac{1}{2}$	$\frac{3}{8}$	$\frac{3}{8}$	$\frac{7}{16}$	$1\frac{1}{16}$	30.7 lbs.
93	5	$2\frac{1}{2}$	$\frac{7}{16}$	$\frac{7}{16}$	$\frac{7}{16}$	$1\frac{1}{16}$	33.0 lbs.
94	4	3	$\frac{3}{8}$	$\frac{7}{16}$	$\frac{3}{8}$	$\frac{3}{8}$	25.9 lbs.
95	4	3	$\frac{5}{16}$	$\frac{5}{16}$	$\frac{5}{16}$	$\frac{5}{8}$	25.25 lbs.
96	4	2	$\frac{5}{16}$	$\frac{5}{16}$	$\frac{3}{8}$	$\frac{9}{16}$	20.4 lbs.
97	3	$3\frac{1}{2}$	$\frac{7}{16}$	$\frac{1}{2}$	$\frac{7}{16}$	$\frac{1}{2}$	28.25 lbs.
98	3	$2\frac{1}{2}$	$\frac{3}{8}$	$\frac{7}{16}$	$\frac{1}{2}$	$\frac{5}{8}$	23.8 lbs.
99	3	$1\frac{1}{2}$	$\frac{1}{4}$	$\frac{1}{4}$	$\frac{1}{4}$	$1\frac{1}{32}$	11.2 lbs.
100	$2\frac{1}{2}$	$1\frac{1}{4}$	$\frac{1}{4}$	$\frac{1}{4}$	$\frac{1}{4}$	$\frac{5}{16}$	9.1 lbs.
101	2	$1\frac{1}{2}$	$\frac{1}{4}$	$\frac{9}{32}$	$\frac{1}{4}$	$\frac{5}{16}$	8.75 lbs.
102	2	1	$\frac{1}{4}$	$\frac{1}{4}$	$\frac{1}{4}$	$\frac{9}{32}$	7.0 lbs.
103	2	$\frac{9}{16}$	$\frac{1}{4}$	$\frac{1}{4}$	$\frac{5}{16}$	$\frac{5}{16}$	5.88 lbs.
104	$2\frac{3}{4}$	$1\frac{3}{4}$	$\frac{5}{16}$	$1\frac{1}{32}$	$\frac{3}{4}$	$\frac{3}{4}$	18.75 lbs.
105	$2\frac{3}{4}$	2	$\frac{5}{16}$	$1\frac{1}{32}$	$\frac{3}{4}$	$\frac{3}{4}$	21.0 lbs.
106	5	$3\frac{1}{2}$	$\frac{1}{2}$	$\frac{1}{2}$	$1\frac{1}{16}$	$\frac{7}{8}$	48.4 lbs.
107	5	4	$\frac{1}{2}$	$\frac{1}{2}$	$\frac{1}{2}$	$1\frac{9}{32}$	44.1 lbs.
108	$2\frac{1}{4}$	$1\frac{9}{16}$	$\frac{1}{4}$	$\frac{1}{4}$	$\frac{5}{16}$	$\frac{5}{16}$	6.5 lbs.
109	4	$4\frac{1}{2}$	$\frac{7}{16}$	$\frac{9}{16}$	$\frac{7}{16}$	$\frac{1}{2}$	38.5 lbs.
110	3	$2\frac{1}{2}$	$\frac{5}{16}$	$\frac{3}{8}$	$\frac{5}{16}$	$\frac{3}{8}$	17.6 lbs.
111	3	$2\frac{1}{2}$	$\frac{3}{8}$	$\frac{7}{16}$	$\frac{7}{16}$	$\frac{3}{8}$	20.6 lbs.

Weights of these sections cannot be varied.

PENCOYD **ANGLE COVERS.**

WEIGHTS PER YARD OF VARIOUS THICKNESSES.

One cubic foot weighing 480 lbs.

CHART NUMBER.	SIZE IN INCHES.	$\frac{1}{8}$	$\frac{3}{16}$	$\frac{1}{4}$	$\frac{5}{16}$	$\frac{3}{8}$	$\frac{7}{16}$	$\frac{1}{2}$	$\frac{9}{16}$	$\frac{5}{8}$
180	3 × 3	....	....	14.3	17.7	21.0	24.2	27.4	30.5	33.5
181	2¾ × 2¾	....	....	13.0	16.1	19.1	22.0	24.9	....	....
182	2½ × 2½	....	....	11.8	14.5	17.2	19.9	22.4	....	....
183	2¼ × 2¼	...	....	10.5	13.0	15.4	17.7	....	....	....
184	2 × 2	....	7.0	9.3	11.4	13.5	....	....	....	....

SIZES OF PENCOYD BAR IRON.

FLATS.

$\frac{7}{8}$	×	$\frac{3}{8}$	inches to	$\frac{3}{4}$	inches.	$2\frac{5}{16}$	× $1\frac{1}{2}$ inches to 2		inches.
1	×	$\frac{1}{4}$	"	$\frac{7}{8}$	"	$2\frac{3}{8}$	× $\frac{5}{8}$	"	$1\frac{7}{8}$	"
$1\frac{1}{16}$	×	$\frac{1}{8}$	"	1	"	$2\frac{1}{2}$	× $\frac{1}{4}$	"	2	"
$1\frac{1}{16}$	×	$\frac{1}{2}$	"	1	"	$2\frac{3}{4}$	× $\frac{1}{4}$	"	$2\frac{1}{4}$	"
$1\frac{1}{8}$	×	$\frac{1}{4}$	"	1	"	$2\frac{13}{16}$	× $1\frac{3}{4}$	"	$2\frac{1}{2}$	"
$1\frac{3}{16}$	×	$\frac{5}{8}$	"	1	"	3	× $\frac{1}{4}$	"	$2\frac{1}{2}$	"
$1\frac{7}{8}$	×	$\frac{1}{2}$	"	1	"	$3\frac{1}{8}$	× $1\frac{3}{4}$	"	3	"
$1\frac{1}{4}$	×	$\frac{1}{4}$	"	1	"	$3\frac{1}{4}$	× $\frac{1}{4}$	"	2	"
$1\frac{5}{16}$	×	$\frac{5}{8}$	"	1	"	$3\frac{1}{2}$	× $\frac{1}{4}$	"	2	"
$1\frac{3}{8}$	×	$\frac{1}{4}$	"	$1\frac{3}{16}$	"	4	× $\frac{3}{8}$	"	$3\frac{1}{2}$	"
$1\frac{13}{32}$	×	$\frac{5}{8}$	"	$1\frac{1}{16}$	"	$4\frac{1}{2}$	× $\frac{1}{4}$	"	2	"
$1\frac{7}{16}$	×	$\frac{1}{8}$	"	$1\frac{3}{16}$	"	5	× $\frac{1}{4}$	"	$3\frac{1}{2}$	"
$1\frac{1}{2}$	×	$\frac{1}{4}$	"	$1\frac{1}{4}$	"	6	× $\frac{1}{4}$	"	3	"
$1\frac{19}{32}$	×	$\frac{5}{8}$	"	$1\frac{5}{16}$	"	7	× $\frac{1}{4}$	"	3	"
$1\frac{5}{8}$	×	$\frac{1}{4}$	"	$1\frac{7}{16}$	"	8	× $\frac{1}{4}$	"	$2\frac{3}{4}$	"
$1\frac{3}{4}$	×	$\frac{1}{4}$	"	$1\frac{1}{2}$	"	9	× $\frac{1}{4}$	"	$2\frac{3}{4}$	"
$1\frac{31}{32}$	×	$\frac{5}{8}$	"	$1\frac{1}{2}$	"	10	× $\frac{1}{4}$	"	$2\frac{1}{4}$	"
2	×	$\frac{1}{4}$	"	$1\frac{7}{8}$	"	11	× $\frac{1}{4}$	"	$2\frac{1}{2}$	"
$2\frac{5}{32}$	× $1\frac{3}{4}$		"	$2\frac{1}{16}$	"	12	× $\frac{1}{4}$	"	$2\frac{1}{2}$	"
$2\frac{1}{4}$	×	$\frac{1}{4}$	"	$1\frac{7}{8}$	"				

SQUARES.

$\frac{1}{2}''$, $\frac{9}{16}''$, $\frac{5}{8}''$, $\frac{11}{16}''$, $\frac{3}{4}''$, $\frac{13}{16}''$, $\frac{7}{8}''$, $\frac{15}{16}''$, $1''$, $1\frac{1}{16}''$, $1\frac{1}{8}''$, $1\frac{3}{16}''$, $1\frac{1}{4}''$, $1\frac{3}{8}''$, $1\frac{1}{2}''$, $1\frac{5}{8}''$, $1\frac{3}{4}''$, $1\frac{7}{8}''$, $2''$, $2\frac{1}{4}''$, $2\frac{1}{2}''$, $2\frac{3}{4}''$, $3''$, $3\frac{1}{4}''$, $3\frac{1}{2}''$, $3\frac{3}{4}''$, $4''$ and $4\frac{1}{4}''$.

ROUNDS.

$\frac{30}{64}''$, $\frac{31}{64}''$, $\frac{1}{2}''$, $\frac{33}{64}''$, $\frac{9}{16}''$, $\frac{37}{64}''$, $\frac{38}{64}''$, $\frac{39}{64}''$, $\frac{5}{8}''$, $\frac{41}{64}''$, $\frac{11}{16}''$, $\frac{46}{64}''$, $\frac{47}{64}''$, $\frac{3}{4}''$, $\frac{49}{64}''$, $\frac{13}{16}''$, $\frac{53}{64}''$, $\frac{54}{64}''$, $\frac{55}{64}''$, $\frac{7}{8}''$, $\frac{57}{64}''$, $\frac{15}{16}''$, $\frac{61}{64}''$, $\frac{62}{64}''$, $\frac{63}{64}''$, $1''$, $1\frac{1}{64}''$, $1\frac{1}{16}''$, $1\frac{5}{64}''$, $1\frac{1}{8}''$, $1\frac{9}{64}''$, $1\frac{3}{16}''$, $1\frac{13}{64}''$, $1\frac{1}{4}''$, $1\frac{17}{64}''$, $1\frac{5}{16}''$, $1\frac{3}{8}''$, $1\frac{7}{16}''$, $1\frac{1}{2}''$, $1\frac{9}{16}''$, $1\frac{5}{8}''$, $1\frac{3}{4}''$, $1\frac{7}{8}''$, $2''$, $2\frac{1}{8}''$, $2\frac{1}{4}''$, $2\frac{3}{8}''$, $2\frac{1}{2}''$, $2\frac{5}{8}''$, $2\frac{3}{4}''$, $2\frac{7}{8}''$, $3''$, $3\frac{1}{8}''$, $3\frac{1}{4}''$, $3\frac{3}{8}''$, $3\frac{1}{2}''$, $3\frac{5}{8}''$, $3\frac{3}{4}''$, $3\frac{7}{8}''$, $4''$, $4\frac{1}{8}''$, $4\frac{1}{4}''$, $4\frac{3}{8}''$, $4\frac{1}{2}''$, $4\frac{5}{8}''$, $4\frac{3}{4}''$, $4\frac{7}{8}''$, $5''$, $5\frac{1}{4}''$, $5\frac{1}{2}''$, $5\frac{3}{4}''$, $6''$, $6\frac{1}{2}''$ and $7''$.

HALF ROUNDS.

$\frac{3}{4}''$, $1''$, $1\frac{1}{4}''$, $1\frac{1}{2}''$, $1\frac{3}{4}''$, $2''$, $2\frac{1}{4}''$, $2\frac{1}{2}''$, $2\frac{3}{4}''$, $3''$, $3\frac{1}{4}''$ and $3\frac{1}{2}''$.

Two grades of iron are manufactured, known respectively as "Pencoyd Refined" and "Pencoyd High Test," the former for all ordinary requirements, the latter for tension members of structures and all purposes where a uniform iron of high ductility is required.

10 × AREA IN INCHES = WEIGHT PER YARD IN LBS.

In any rolled section of wrought iron, the weight in lbs. per yard, is precisely equal to ten times its sectional area in square inches.

Consequently, either value being known, the other can be instantly obtained.

AXLES.

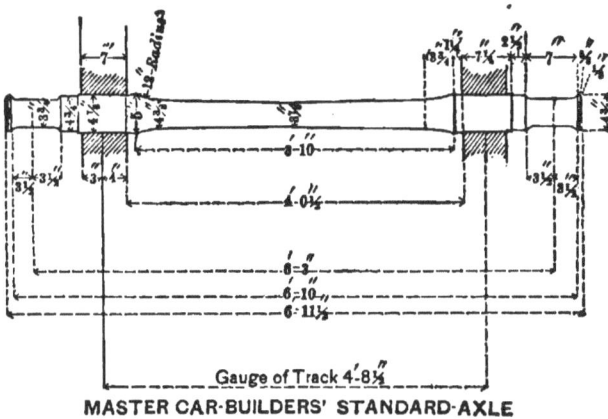

MASTER CAR-BUILDERS' STANDARD-AXLE

Hammered or rolled axles of iron or steel, centred and straightened with journals forged or rough-turned, made to conform to specifications and tests.

STRUCTURAL WORK.

The fitting, punching, and riveting of structural work executed, and iron castings furnished to order.

MISCELLANEOUS SHAPES.

CAR BUILDERS' CHANNEL.

Chart No. 33.

Weight per yard = 50 to 55 lbs.

TEN-INCH BULB PLATE.

Chart No. 68.

Weight per yard = 62 lbs.

MINERS' TRACK RAIL.

Chart No. 190.

Weight per yard = 25 lbs.

SPLICE BAR FOR MINERS' TRACK RAIL.

Chart No. 191.

Weight per yard = 5.2 lbs.

SLOT RAIL FOR CABLE ROAD.

Chart No. 192.

Weight per yard = 26 lbs.

HALF OVALS.

Chart No. 193 = 4.3 lbs. per yd.
Chart No. 194 = 4.8 lbs. per yd.

Channel Rail.   Chart No. 195 = 3.5 lbs. per yard.

GROOVED BARS.

Chart No. 196 =   8.4 to 14.7 lbs. per yard.
       "   197 = 13 5 to 21.0 lbs. per yard.
       "   198 = 20.9 to 34 5 lbs. per yard.

STRENGTH OF WROUGHT IRON.

The tensile strength of rolled iron varies according to the quality of the material, the mode of manufacture, and the sectional area of the bar. In general terms the ordinary sizes of bars of good material may be accepted as having an ultimate tensile strength of 50,000 lbs. per square inch of section, an elastic limit of 30,000 lbs., and will stretch 20 per cent. in a length of 8 inches when tested up to rupture.

It is, however, as easy to produce the smaller sizes yielding results 10 per cent. higher than the above, as it is difficult to make the largest sections with a limit 10 per cent. below the same figures.

Dividing rolled iron into three classes according to its sectional area, we have:

I.—Bars not exceeding 1¼ square inches area.

II.—Bars from 1¼ to 4 square inches area.

III.—Bars from 4 to 8 square inches area.

For which experiments give the following figures as average results.

CLASS.	TENSILE STRENGTH PER SQ. INCH.	ELASTIC LIMIT PER SQ. INCH.	ELONGATION IN 8 INCHES.
I.	53,000 lbs.	33,000 lbs.	25 per cent.
II.	50,000 "	30,000 "	20 " "
III.	48,000 "	28,000 "	18 " "

These, however, are only general conclusions, as much depends on the shape of the section, the method of rolling, and the reduction of area from the pile to the finished bar.

The following tensile tests are actual averages taken from our records, and were made on specimens cut from bars of the sizes and shapes given, and intended for use in bridges, and to conform to the specifications of the leading railroad companies.

Size and Shape of Bar.	Ultimate strength in lbs. Per Square Inch.	Elastic Limit in lbs. Per Square Inch.	Per Cent. of Elongation in 8 inches.	Per Cent. of Reduction of Fractured Area.	
One-inch rounds.	52,210	32,150	26	39	....
Two-inch rounds.	50,935	31,800	19.8	....	....
Four-inch rounds.	48,220	26,640	18	....	....
Four-inch flats.	51,000	30,000	20.7	31	$\frac{3}{4}$ to $1\frac{1}{2}$ in. thick.
Eight-inch flats.	49,500	31,500	16	....	$\frac{1}{2}$ inches thick.
Twelve-inch flats.	49,080	31,560	15.5	....	$\frac{3}{4}$ inches thick.
Three-inch angles.	49,000	30,500	17	30	....
Six-inch angles.	49,160	30,150	18.1	....	....
Flanges of beams.	51,840	31,560	20.1	....	....
Webs of beams.	50,130	30,150	17.7	....	....

COMPRESSION.

The power of wrought iron to resist compression is usually taken as equal to its tensile strength. In the form of flanges for solid beams, this property is exerted to its full capacity, as the adjacent portion of the material in tension sustains the portions in compression from buckling, even when the length of the beam becomes very considerable. But in the form of struts and columns, when the piece becomes of considerable length in proportion to its cross-section, failure occurs by bending, or combined bending and crushing. (See article on Struts.) Judging from many experiments we have made on bars secured from bending under compressive stress, the elastic limit in compression is a little lower than in tension, but the former not so clearly defined as the latter; practically they may be considered as equal.

These results were derived from small sections; in large sections there may be more equality, as some experiments hereafter described would denote.

With pressures varying from 25,000 to 35,000 lbs. per square inch, the elastic limit is attained. With 50,000 lbs. per square inch a permanent reduction of $2\frac{1}{2}$ per cent. of the length is produced; with 75,000 lbs. a reduction of 6 per cent., and with

100,000 lbs. per square inch the permanent reduction of length is about 8 per cent. These results have a wide range of variation, but the figures are the averages of several experiments.

ELASTICITY OF ROLLED IRON.

The elasticity of wrought iron, or its ratio of change of length under stress below the elastic limit, varies more extensively than any other property of rolled iron. Experiment shows a variation of over 100 per cent. in extreme cases.

The modulus of elasticity is an imaginary load, which, supposing the material to be perfectly elastic, would cause the iron to double its length under tension, or to shorten its length one-half under compression, and return to its original length when released from stress. This modulus is usually assumed at 29,-000,000 lbs. In large sections of properly prepared material the tensile elasticity probably averages a little over this, and the compressive elasticity a little below it.

The following results of the tests for comparative elasticity in tension and compression, will serve to illustrate the irregularity of the elasticity; also, see tests of iron and steel cut from beams, given hereafter.

Two pieces of ¾-inch square iron cut from same bar.
Measured length of each specimen = 12 inches.
Area of each specimen = .556 square inch.
Pressures in lbs. ; change of length in inches.

TENSILE TEST.			COMPRESSIVE TEST.		
	Elongations.			Reduction of length.	
Pressure per sq. inch.	Load on.	Load off.	Pressure per sq. inch.	Load on.	Load off.
5,000	.002	.000	5,000	.002	.000
10,000	.0045	.000	10,000	.0035	.000
15,000	.0065	.000	15,000	.005	.000
20,000	.0085	.000	20,000	.006	.000
22,000	.010	.000	22,000	.007	.000
24,000	.0105	.000	24,000	.008	.000
26,000	.0115	.000	26,000	.009	.000
28,000	.012	.000	28,000	.0095	.000
30,000	.013	.000	30,000	.010	.000
32,000	.0135	.000	32,000	.011	.000
34,000	.0145	.000	34,000	.020	.0035
36,000	.0155	.001	36,000	.023	.0045
38,000	.1715	.1495	38,000	.027	.010
40,000	.3835	.3605	40,000	.107	.089
50,000	1.326	1.2945	50,000	.272	.246
53,820	3.093	......	60,000	.464	.435
			70,000	.671	.639
			80,000	.845	.814
			90,000	1.074	1.042

Specimen broke with 53,820
lbs. per square inch.
Stretched 3.093 in 12 in.
" 2.187 in 8 in.
" 27.3 per cent. in 8 in.

Fractured area = .3364

Modulus of elasticity
= 27,420,000 lbs.

Modulus of elasticity
= 35,300,000 lbs.

Two pieces of ¾-inch round iron cut from same bar.
Measured length of each specimen = 12 inches.
Area of each specimen = .449 square inch.
Pressure in lbs.; change of length in inches.

TENSILE TEST.			COMPRESSION TEST.		
	Elongations.			Reduction of length.	
Pressure per sq. inch.	Load on.	Load off.	Pressure per sq. inch.	Load on.	Load off.
5,000	.002	.000	5,000	.002	.000
10,000	.004	.000	10,000	.005	.000
15,000	.006	.000	15,000 .	.007	.000
20,000	.008	.000	20,000	.010	.000
22,000	.009	.000	22,000	.011	.001
24,000	.010	.000	24,000	.012	.002
26,000	.0105	.000	26,000	.013	.003
28,000	.011	.000	28,000	.015	.0045
30,000	.013	.000	30,000	.0215	.0065
32,000	.014	.000	32,000	.0225	.007
34,000	.015	.002	34,000	.0275	.009
36,000	.022	.007	36,000	.040	.019
38,000	.416	.399	38,000	.052	.036
40,000	.544	.523	40,000	.133	.114
50,000	1.740	1.707	50,000	.304	.283
51,600	2.468	.....	60,000	.427	.402
			70,000	.546	.521
			80,000	.663	.635
			90,000	.773	.742
			100,000	.896	.862

Specimen broke with 51,600
lbs. per square inch.
Stretched 2.468 in 12 in.
    "   1.81 in 8 in.
    "   22.6 per cent. in 8 in.

Fractured area = .297 sq. in.

Modulus of elasticity
    = 29,400,000 lbs.

Modulus of elasticity
    =24,490,000 lbs.

A series of tests was made on the United States Government testing machine at Watertown Arsenal, on the full-sized bars, of which the following is a condensed average.

TENSILE TESTS.

MODE OF MANU-FACTURE.	Section of bars.	Ultimate tenacity in lbs. per sq. in.	Elastic limit in lbs. per sq. in.	Reduced area at fracture, per cent.	Modulus of Elasticity.
Single rolled..	3 × 1	50,600	28,600	29	28,200,C00
Double rolled..	3 × 1	52,500	30,100	32	27,885,000
Single rolled..	5 × 1¼	49,800	26,100	21	27,930,000
Double rolled..	5 × 1¼	51,000	27,200	28	28,920,000

The "single and double rolled" means the number of workings from the puddled bar.

A number of experiments on large columns with the same machine gave the following results,—also the tensile results, for the iron used in the construction of the columns.

	ELASTIC LIMIT.	MODULUS OF ELASTICITY.
Wrought iron in compression	27,500	29,C00,000
Wrought iron in tension..........	31,600	£9,100,000

The modulus of transverse elasticity as applied to our tables of deflections is taken at 26,000,000 lbs. It is a hypothetical quantity, derived by means of formulæ, which are given elsewhere, and which assume that the resistances to tension and compression are equal, and that the successive fibres of iron, from the neutral axis outward act independently of each other, neither of which statements are correct in fact.

It is probable that this modulus, with the same material, will vary with each change of section, and possibly also with changes of length, and conditions of load.

SHEARING.

Under the conditions that shearing stresses are usually applied in structures, the shearing strength of wrought iron is about eight-tenths of the tensile, viz., 40,000 lbs. per square inch of section. But when subjected to the action of properly prepared cutting knives, the resistance to shearing is much less than this.

TORSION.

The resistance to twisting is proportional to the cube of the diameter. When the shearing strength is known, the torsional strength of any round shaft can be determined as follows : $T = 1.57 \, sr^3$. $r =$ radius of shaft in inches. $s =$ shearing strength in lbs. per square inch. $T =$ the torsional moment in inch lbs., or the force in lbs. multiplied by the leverage in inches with which it acts.

In practice, however, torsion is usually accompanied by bending stresses, which must be always considered when determining the proportions of shafts. See article on Shafting, page 170.

STRUCTURAL STEEL.

The various grades of steel used in structures possess such an extended range of physical properties that it is impossible to present as definite a basis for strength, stiffness, etc., as can be given for wrought iron.

The character of the material is largely determined by its combination, in minute proportions, with various substances, the most important of which is carbon.

As a general rule the greater the percentage of carbon in the steel, the higher will be its tensile strength and the lower its ductility. The following list exhibits the average tensile resistances for steels having given proportions of carbon:

PERCENTAGE OF CARBON.	TENSILE STRENGTH IN POUNDS PER SQUARE INCH.		DUCTILITY.
	ULTIMATE TENACITY.	ELASTIC LIMIT.	ULTIMATE ELONGATION IN 8 INCHES.
.10	60000	36000	26 per cent.
.15	66000	40000	24 "
.20	74000	45000	22 "
.25	82000	50000	20 "
.30	90000	55000	18 "
.35	100000	60000	16 "
.40	110000	65000	14 "

These figures, however, are only approximate, as much depends on the quality of the steel, and also the extent to which it has been worked in the rolling process.

The grades below .15 per cent. carbon are known conventionally as "mild steels," owing to their high ductility and to their possessing but very moderate hardening properties when chilled in water from a red heat.

The mild steel has also superior welding properties, as compared with hard steel, and will endure higher heat without injury.

Steel whose carbon ratio does not exceed .10 per cent. should be capable of doubling flat without fracture, when chilled in the coldest water from a red heat.

Steel of .12 carbon should endure similar treatment when chilled in water of 80° F.

When the carbon percentage is .15 the steel should be capable of bending at least 90°, over a curve whose radius is three or four times the thickness of the specimen operated upon, and after being chilled from a red heat in water of 80° F.

Steel having .35 to .40 per cent. carbon, will usually harden sufficiently to cut soft iron, and maintain an edge.

There is much variation from the aforesaid hardening properties in different qualities of steel, as much depends on the influence of other hardening agents besides carbon.

The modern tendency is to limit the use of steel for structural purposes to the milder grades of the material. For steel in steamships the United States Government specifies as follows : "Steel to have an ultimate tensile strength of not less than 60,000 lbs. per square inch, and a ductility of not less than 25 per cent. in 8 inches. The test piece to be heated to a cherry-red and chilled in water at a temperature of 82° F. After this it must be capable of bending double flat under the hammer without cracking." It requires about .11 to .12 carbon steel to endure this test.

"Lloyd's" rules require the steel to have an ultimate tenacity of not less than 60,000, or not over 70,000 lbs. per square inch, with an elongation of at least 16 per cent. in 8 inches. This steel, when heated to redness and chilled in water of 82° F., must bend double without fracture around a curve of which the diameter is not more than three times the thickness of the piece tested. For a cold test without hardening, the material must be capable of doubling flat and bending backward without fracture.

Angles and beams for ship-frames may have a tenacity of 74,000 lbs., providing the bending tests are satisfactory, and the welding property is unimpaired. It requires about .12 to .14 carbon steel to meet these specifications.

We have made numerous experiments on steel of several grades and in various forms, but the resistance under stress is so uncertain that a fair statement of its physical properties cannot be satisfactorily given until an exhaustive series of experiments has been made on material of definite composition.

We present the average results of experiments on the strength and elasticity of "mild" and "hard" steel, also the comparative resistance of these materials in the form of struts. The "mild steel" had an average carbon ratio of .12 per cent., and the "hard steel" an average carbon ratio of .36 per cent. The average strength and elasticity of wrought iron is inserted for the purpose of exhibiting the characteristics of the steel and iron. As in the case of the steel, the several values given for iron are the results of a few special experiments.

| MATERIAL. | TENSILE STRENGTH IN LBS. PER SQUARE INCH. | | DUCTILITY. | MODULUS OF ELASTICITY IN LBS. |
	ULTIMATE TENACITY.	ELASTIC LIMIT.	ELONGATION IN 8 INCHES.	
Iron.........	51000	31000	19 per cent.	28400000
Mild steel....	64000	39000	24 "	29300000
Hard steel...	100000	56700	18 "	29280000

From the same material the following results for compression were obtained.

COMPRESSIVE RESISTANCE.

MATERIAL.	ELASTIC LIMIT IN LBS. PER SQUARE INCH.	MODULUS OF ELASTICITY.
Iron	29500	27090000
Mild steel	37400	24760000
Hard steel...........	55700	24570000

TRANSVERSE STRENGTH.

A series of experiments was made on the transverse strength and elasticity of round bars from 3 to 4 inches in diameter, and flanged beams varying from 3 to 12 inches deep, and from 3 feet to 20 feet in length. For the purpose of making a compact exhibit of the resistance of beams of various lengths and cross sections, the results of the experiments were condensed to the method of the ensuing table, in which

R = the modulus of maximum resistance.
R_1 = the modulus of resistance at the elastic limit.
E = the modulus of transverse elasticity.

$$R \text{ or } R_1 = \frac{\text{bending moment} \times \text{depth of beam}}{2 \times \text{Inertia}}.$$

$$E = \frac{\text{Weight} \times \text{cube of length}}{48 \times \text{Inertia} \times \text{deflection}}.$$

The ultimate resistance was taken at that stage of the experiment where increase of deflection occurred without increase of load.

MATERIAL.	R	R_1	E
Iron	44700 lbs.	31000 lbs.	27600000 lbs.
Mild steel	52800 "	39500 "	29700000 "
Hard steel..........	80200 "	54500 "	27200000 "

As is well known, the elasticity of iron is so variable and uncertain, that no definite value can be assigned to it except by taking the averages of numerous experiments. Steel possesses the same uncertain elasticity, especially under transverse and compressive stresses.

The elastic moduli in tension varied from 27 to 33 millions of pounds, in compression from 21 to 33 millions, and transversely the modulus of elasticity varied from 23 to 33 millions of pounds.

It is probable that there is not much difference on the whole between the transverse elasticity of iron and either grade of steel; if any difference at all exists, the steel probably has the advantage in stiffness, and the experiments indicate that the mild steel, if anything, is stiffer than the hard steel, the reverse of what is popularly supposed to be the case.

STEEL BEAMS.

The experiments demonstrate that the transverse resistance of steel of different grades maintains a ratio practically uniform with the tenacities of the different steels.   Consequently when steel of known tensile strength is used in beams, the absolute strength of the beam may be obtained from our rules and tables for iron by increasing the results in the proportion of the increased tenacity of the particular steel used over that of iron. The percentage of increase for good qualities of steel, will be about as follows :

CARBON PERCENTAGE.	INCREASED STRENGTH OF STEEL OVER WROUGHT IRON BEAMS.
.10	20 per cent.
.15	35 "
.20	50 "
.25	65 "
.30	80 "

The experiments do not show that steel of any grade is stiffer under working loads than wrought iron.   Therefore beams of either steel or wrought iron having uniform lengths and cross sections will deflect uniformly under equal loads, below the elastic limit of wrought iron, and our tables of deflections for iron beams as given hereafter, will apply also to steel.

STEEL SHAFTING.

When absolute strength irrespective of stiffness is alone considered, steel probably possesses a torsional strength exceeding that of iron about in the ratio of the respective tenacities of the two metals. Therefore, when designing shafting under such conditions, our formulæ for iron shafting can be used, substituting a shearing resistance equal to $\frac{3}{4}$ of the tensile strength of the steel, in place of that given for iron in the article on Shafting. But in the large majority of cases the usefulness of shafting is determined by its transverse stiffness, irrespective of its ultimate torsional strength.

As in this respect the advantage of steel over iron is very questionable, it will be found necessary to use the same dimensions of steel shafts as determined by our rules for wrought iron.

STEEL STRUTS.

The experiments on direct compression prove that the elastic limits of steel, as of iron, under stresses of tension and compression, are about equal.

Consequently for the shortest struts, where failure results from the effects of direct compression, the tensile resistances of steel and iron serve as a comparative measure of the strut resistance of the two materials.

But as the strut is increased in length, and failure results from lateral flexure before the compressive limit of elasticity is attained, then the transverse elasticity of the material becomes a factor of increasing importance in determining the strut resistance.

As in this respect the steel possesses little advantage, if any, over iron, the tendency will be for struts of steel and iron as the length is increased to approximate toward equality of resistance. This equality with iron will be attained, first by the mildest steel, and latest by the hardest steel.

The results of many experiments we have made seem to demonstrate that this equality of strut resistance is practically attained between iron and mild steel, when the ratio of length to least radius of gyration of cross section is about 200 to 1. In

the case of the harder steels, practical equality of resistance
would probably be reached at some higher but unknown ratio of
length to section.

We give a table exhibiting the comparative resistances per
square inch of section for flat-ended struts of iron, mild steel,
and hard steel, and for further particulars of the subject refer
to the article on Struts, given hereafter.

It is quite probable that grades of steel intermediate between
those denoted in the table will offer intermediate resistance as
struts, in the ratio of their percentage of carbon, other elements
remaining the same.

SPECIFIC GRAVITY.

The specific gravity of steel and iron varies according to the
purity of the metal, and also to the degree of condensation im-
parted by the rolling process.

As a rule the mild steel has a higher specific gravity than
hard steel, and both are denser than iron. A number of tests
we have made for specific gravity show rolled bars of mild steel
to vary from 7.84 to 7.83, and hard steel from 7.81 to 7.85
specific gravity. Ordinary iron bars will vary from 7.6 to 7.8.

In the form of beams and large rolled sections generally, the
following figures may be accepted as a fair average.

Material.	Weight per cubic foot.	Weight per cubic inch.
Mild Steel	489.0 lbs.	.283 lb.
Hard Steel	486.6 "	.2815 "
Iron	478.3 "	.2768 "

Or for the same sectional areas, the excess in weight over iron
will be, for mild steel 2.24 per cent. and for hard steel 1.7 per
per cent.

FLAT-ENDED STRUTS.

ULTIMATE RESISTANCE IN POUNDS PER SQUARE INCH OF SECTION.

LENGTH DIVIDED BY LEAST RADIUS OF GYRATION.	IRON.	MILD STEEL. .12 CARBON.	HARD STEEL. .36 CARBON.
20	46000	70000	100000
30	43000	51000	74000
40	40000	46000	62000
50	38000	44000	60000
60	36000	42000	58000
70	34000	40000	55500
80	32000	38000	53000
90	30900	36000	49700
100	29800	34000	46500
110	28000	32000	43200
120	26300	30000	40000
130	24900	28000	36700
140	23500	26000	33500
150	21750	24000	30700
160	20000	22000	28000
170	18400	20000	25500
180	16800	18000	23000
190	15650	16200	21000
200	14500	14800	19000
210	13600 ·	13600	17200
220	12700	12700	15500
230	11950	11950	14400
240	11200	11200	13400
250	10500	10500	12400
260	9800	9800	11500
270	9150	9150	10600
280	8500	8500	9700
290	7850	7850	9000
300	7200	7200	8500

RESISTANCE TO BENDING.

When wrought-iron beams are subjected to bending stresses, the resulting deflections increase nearly in direct proportion to the increase of load, up to the limit of elasticity of the iron. Slight permanent sets can be observed in the beam before the elastic limit is reached, just as similar sets are obtained in longitudinal tests. After the elastic limit is passed, the deflections increase in a greater ratio than the loads, and clearly defined permanent sets occur, until another stage in the experiment is reached, when the beam shows increasing deflection without any increase of load. At this point the element of time becomes an important factor. The load can be very slowly increased, without the record of stress showing increase, but if the load is freely applied, the recorded stress may be very considerably augmented. It is probable that if the load was left long enough on the beam at this stage of the experiment entire failure would ensue.

We call this point, which can generally be very clearly observed, the "ultimate resistance" of the beams, and whenever such terms as "ultimate load," "breaking load," etc., are used in connection with bending stresses, this is the load referred to. The stress at the elastic limit bears no such fixed relation to the ultimate stress as can generally be observed in tensile tests. The length of the beam, and probably other conditions, such as position of load, etc., become factors in determining the ratio, which in the absence of complete experiments cannot be decided.

MODULUS OF RUPTURE.

If the material of a beam offered equal resistances to tension and compression, and if the fibres acted independently of each other in effecting this resistance, then the maximum fibre stresses, which occur at the top and bottom of the beam, could be readily calculated as follows:

For any rectangular section loaded in the middle $S = \dfrac{3\,w\,l}{2\,b\,d^2}$;

for a beam 1 inch square and 12 inches long, $S = 18\,W$, or in general terms for any symmetrical beam, under any condition of load, $S = \dfrac{M\,d}{2\,I}$.

$S=$ maximum fibre stress. $w=$ load.

$b=$ breadth of beam. $l=$ length of beam.

$d=$ depth of beam. $M=$ bending moment.

$I=$ moment of inertia about the neutral axis at right angles to the direction of pressure.

But, as previously stated, neither of these usually assumed conditions exist.

It seems probable that the fibres nearer the axis, by means of lateral adhesion, relieve the outer fibres from a portion of the stress which the usually accepted theory indicates, and consequently have their own portion of the theoretical stress correspondingly increased. It is therefore necessary to abandon the deceptive term of "maximum fibre stress," and substitute a " modulus " determined by means of the foregoing formulæ.

This modulus will vary for varying cross-sections, and recent experiments make it seem probable that it will vary with the length of beam, etc.

The average of a large number of experiments on standard flanged beams give an ultimate modulus of 42,000 lbs. On solid rectangular sections the modulus will run higher, or from 45,000 to 50,000 lbs.

We adopt 42,000 as the modulus for ultimate transverse strength of I beams. All our tables are calculated by taking $S=14,000$, or one-third of the ultimate strength of the beam.

LIMITS FOR THE SAFE LOAD.

Inasmuch as there is a great diversity in published tables of safe loads for beams, every one must judge for himself what proportion of the elastic strength of the beam will best suit his purpose.

The character of the load must be considered, and the mode of application of the same. If the load is suddenly applied, especially if accompanied by impact, the dynamic stresses resulting therefrom will not be expressed by formulæ which are derived from static considerations alone. Freedom from vibration or excessive deflection have usually to be provided for, or the beam may be of considerable length without lateral support. In many such cases it may be necessary to take one-fourth or one-fifth of the ultimate strength of the beam as the working basis, instead of

3

one-third, as given in our tables, which we give as the "greatest safe loads."

We have every confidence in the accuracy of the tables, as the results of a number of careful tests we have recently made show that very rarely does the ultimate strength of the beam fall below the limits we have given, and in some instances it considerably exceeds those limits.

We have in our own service beams that are continually subjected to much higher bending stresses than would be assigned to them by our tables without any evidence of a want of stability.

FACTOR OF SAFETY.

For factors of safety the following table will give results in harmony with good practice.

CHARACTER OF STRESS.	GREATEST SAFE LOAD.
Quiescent load, subject to little or no vibration as in light roofs, etc.	$\frac{1}{3}$ of ultimate.
Fluctuating loads causing vibration, but no sudden application of the maximum load. Such as lateral bracing of bridges, roofs carrying shafting, etc.	$\frac{1}{4}$ of ultimate.
When maximum loads are suddenly applied.	$\frac{1}{5}$ of ultimate.
When maximum stresses are suddenly reversed in direction.	$\frac{1}{6}$ of ultimate.

UNSYMMETRICAL BEAMS.

When beams have not an identical cross-section above and below the neutral axis, as in Deck Beams, Tees, Angles, etc., experiment shows no substantial difference in either the strength or stiffness of the beams, whether the greatest flange is in tension or compression, up to or nearly to the elastic limit. When the least flange is in compression the elastic limit ranges a little higher than when it is in tension, and in the former case, after the elastic limit is passed, the beam generally exhibits much less deflection and higher ultimate resistance than when loaded with the least flange in tension. This is probably due to the high resistance of wrought iron to crushing after the elastic limit is passed.

There are some exceptions to this, as in the case of very long beams that present no adequate resistance to lateral flexure, but as such cases are outside the bounds of good practice they require no further notice. The authoritative formulæ most generally accepted are based upon a maximum fibre stress obtained as follows: $S = \dfrac{M\,d}{I}$. $M =$ bending moment. $d =$ distance from neutral axis to farthest edge of section. $I =$ moment of inertia about the axis passing through the centre of gravity at right angles to direction of pressure. This does not give results in harmony with experiments, except by taking S as a modulus, whose value would not agree with that used for symmetrical beams, and whose value would have to be derived by experiments for differing cross-sections. By taking the moments of inertia above and below an axis so located that the forces producing tension and compression are in equilibrium, and using the modulus, $S = 42,000$, as in symmetrical beams, results harmonizing with experiments are obtained.

But, for simplicity, we have adopted the following methods for calculating the safe load, which, though incorrect in principle, yet give correct results for the particular sections referred to.

$$\text{Deck Beams } \frac{M\,d}{2\,I} = S = 42,000.$$

Tees and Angles of equal legs and uniform thickness. $\left. \right\} \dfrac{M\,d}{2\,I} = S = 45,000.$

Notation as for equal flanged beams.

PENCOYD BEAMS.

GREATEST SAFE LOADS.

The following tables for I beams, channels, and deck beams give the greatest safe loads in net tons, evenly distributed over the beams, and including the weight of beam itself.

These loads are one-third ($\frac{1}{3}$) of the ultimate strength of the beams, and are correct for the corresponding sectional areas given. The several values are obtained by the methods described on page 88, and have been confirmed by numerous experiments. The beams, if of considerable length, are supposed to be braced horizontally, and it is safest to limit the application of the tabular loads to beams whose length between lateral supports does not exceed twenty times the flange width.

Our experience has been that a beam without lateral support is much more stable than is commonly supposed. In an open webbed beam, the top flange acts as a simple strut, and is liable to lateral flexure when the unsupported length is considerable. But in a solid beam the parts in tension sustain the parts in compression rigidly, and prevent the buckling which would otherwise occur.

A number of careful experiments have shown a reduction of about one-third of the normal modulus of rupture when the length of the beam becomes 80 times its flange width. But as the long beam may suffer if exposed to accidental cross strains, we recommend the greatest safe load to be reduced in such a ratio for long beams that when the length is seventy times the flange width the greatest safe loads will be reduced one-half. This will give safe loads, corresponding to given lengths as follows:

BEAMS WITHOUT LATERAL SUPPORT.

LENGTH OF BEAM.	PROPORTION OF TABULAR LOAD FORMING GREATEST SAFE LOAD.
20 times flange width.	Whole tabular load.
30 " " "	$\frac{9}{10}$ " "
40 " " "	$\frac{8}{10}$ " "
50 " " "	$\frac{7}{10}$ " "
60 " " "	$\frac{6}{10}$ " "
70 " " "	$\frac{5}{10}$ " "

The safe loads for any other length, not given in the tables can readily be found by simple proportion, remembering if the span is very short to limit the load to that given in col. xiv, pages 93-97, headed "Maximum load in tons." If beams of any sectional area not given in the tables are used, the strength can be found as described on page 106, or a close approximation to the same by the rule on page 69.

DEFLECTION.

Inasmuch as the elasticity of iron and steel is very variable and uncertain, the tabular deflections are given as the nearest probable, and are obtained as described on page 89.

The tabular deflections correspond to the given loads evenly distributed, and apply to any sectional area for each size of beams respectively, when the corresponding loads bear a uniform ratio to the strength of the beam.

The greatest safe load in the middle of the beam is exactly one-half ($\frac{1}{2}$) of the distributed load, and the deflection for the former will be eight-tenths ($\frac{8}{10}$) of the deflection corresponding to the distributed load as given in the tables. If the load is placed out of centre on the beam, it will bear the same ratio to the load at the centre that the square of half the span bears to the product of the segments of the beam formed by the position of the load.

Example.—A 15-inch 200 lb. **I** beam, 16 feet between supports, will safely carry an evenly distributed load (by the tables) of 26.5 tons, and deflect under same .27 inches. The greatest safe load in the middle will be one-half the above, viz., 13.25 tons, and the resulting deflection $\frac{8}{10}$ of the former, or .22 inches.

If the weight is concentrated 3 feet out of centre, or 5 feet and 11 feet from the ends, then the square of half the span being 64, and the product of the segments being 55, the greatest safe load will be $\dfrac{13.25 \times 64}{55} = 15.4$ tons.

If a beam of above size and length is used without any lateral support, reduce the safe load in the ratio aforesaid. Thus the flange is $5\frac{3}{4}$ inches wide, and the length 33 times this; therefore the greatest safe load will be a little less than $\frac{9}{10}$ of the results in the example.

If the beam is exposed to much vibration, or the action of moving loads, etc., reduce the tabular loads, as previously described on page 34.

For beams of other character than described, the greatest safe loads and corresponding deflections will bear the following ratios to the tabulated loads, for the same lengths of beams :

CHARACTER OF BEAM.	GREATEST SAFE LOAD.	DEFLECTION.
Fixed at one end, with the load concentrated at the other end.	One-eighth ($\frac{1}{8}$) part of the tabular load.	Three and one-fifth ($3\frac{1}{5}$) times the tabular deflection.
Fixed at one end, with the load uniformly distributed.	One-fourth ($\frac{1}{4}$) part of the tabular load.	Two and two-fifths ($2\frac{2}{5}$) times the tabular deflection.
Rigidly fixed at both ends, with a load in the middle of beam.	Same as the tabular load.	Four-tenths ($\frac{4}{10}$) of the tabular deflection.
Rigidly fixed at both ends, with the load uniformly distributed.	One and one-half ($1\frac{1}{2}$) times the tabular load.	One-sixth ($\frac{1}{6}$) of the tabular deflection.
Continuous beam loaded in middle.	Same as the tabular load.	Four-tenths ($\frac{4}{10}$) of the tabular deflection.
Continuous beam load uniformly distributed.	One and one-half ($1\frac{1}{2}$) times the tabular load.	One-sixth ($\frac{1}{6}$) of the tabular deflection.

BEAMS WITH FIXED ENDS.

It is necessary to bear in mind the distinction between ends " rigidly fixed " and ends simply " supported," the latter being the class contemplated in all our tables of safe loads. By "rigidly fixed," as denoted in the previous table, we mean that the beam must be so securely fastened at both ends, by being built into solid masonry, or so firmly attached to an adjacent structure, that the connection would not be severed if the beam was exposed to its ultimate load. In this case, the beam is of the same character as if continuous over several supports, or as if consisting of two cantilevers, the space between whose ends was spanned by a separate beam.

CONTINUOUS BEAMS.

If a beam is continuous over several supports, and is equally loaded on each span, the greatest safe loads and the resulting deflections on any intermediate span will be as given in the preceding table. But the end spans of such a beam, being only semi-continuous, must be either of a shorter span than the intermediates, or if of the same length, the load must be diminished. See "Continuous Beams," page 75.

LIMIT FOR DEFLECTION.

It is considered good practice in the case of plastered ceilings, or in other circumstances where undue deflection may be prejudicial, to proportion beams so that their deflection will not exceed $\frac{1}{30}$ of an inch per foot of span, or $\frac{1}{360}$ part of the span. A heavy black line is marked across, or partly across, each page. All beams below these lines will deflect in excess of this limit; those above the line are safe to use.

PENCOYD   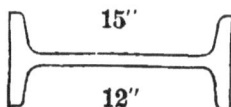   BEAMS.

15″ / 12″

Maximum and Minimum sections of each shape.
Greatest safe load in Net Tons evenly distributed, including beam itself.
Deflections in inches corresponding to given loads for each size of beam.
For a load in middle of beam, allow one-half the tabular figures.
Deflection for latter load will be $\frac{8}{10}$ of the tabular deflection.

LENGTH OF SPAN IN FEET	CHART NUMBER 1	1	2	2	DEFLECTIONS FOR 15″ BEAMS	3	3	4	4	DEFLECTIONS FOR 12″ BEAMS
SIZE OF BEAM IN INCHES	15″	15″	15″	15″		12″	12″	12″	12″	
WT. PER YD. IN LBS.	233	200	201	145		194	168	163	120	
MOMENT OF INERTIA	743.6	682.1	626.6	521.2		403.5	372.0	324.6	272.9	
	GREATEST SAFE LOAD					GREATEST SAFE LOAD				
10	46.29	42.44	38.96	22.10	.11	31.36	28.93	25.24	21.22	.13
11	42.08	38.58	35.42	22.10	.13	28.51	26.30	22.95	19.29	.16
12	38.57	35.37	32.47	22.10	.15	26.13	24.11	21.03	17.69	.19
13	35.61	32.65	29.97	22.10	.18	24.12	22.25	19.42	16.32	.22
14	33.06	30.31	27.83	22.10	.21	22.40	20.67	18.03	15.16	.26
15	30.80	28.29	25.97	21.62	.24	20.91	19.29	16.83	14.15	.30
16	28.93	26.53	24.35	20.27	.27	19.60	18.08	15.76	13.26	.34
17	27.23	24.96	22.92	19.08	.30	18.45	17.02	14.85	12.48	.38
18	25.72	23.58	21.64	18.02	.34	17.42	16.07	14.02	11.79	.43
19	24.36	22.34	20.51	17.07	.38	16.51	15.23	13.28	11.17	.48
20	23.14	21.22	19.48	16.21	.42	15.68	14.47	12.62	10.61	.53
21	22.04	20.21	18.55	15.44	.46	14.93	13.78	12.02	10.11	.58
22	21.04	19.29	17.71	14.74	.51	14.25	13.15	11.47	9.65	.64
23	20.13	18.45	16.94	14.10	.56	13.63	12.58	10.97	9.23	.70
24	19.29	17.68	16.23	13.51	.61	13.07	12.05	10.52	8.84	.77
25	18.52	16.98	15.58	12.97	.66	12.54	11.57	10.10	8.49	.83
26	17.80	16.32	14.99	12.47	.72	12.06	11.13	9.71	8.16	.90
27	17.14	15.72	14.43	12.01	.77	11.61	10.72	9.35	7.86	.97
28	16.53	15.16	13.91	11.58	.83	11.20	10.33	9.01	7.58	1.05
29	15.96	14.63	13.43	11.16	.89	10.81	9.98	8.70	7.32	1.12
30	15.43	14.15	12.99	10.81	.95	10.45	9.64	8.41	7.07	1.20
31	14.93	13.69	12.57	10.46	1.02	10.12	9.33	8.14	6.85	1.28
32	14.47	13.26	12.17	10.13	1.09	9.80	9.04	7.89	6.63	1.36
33	14.03	12.86	11.81	9.83	1.16	9.50	8.76	7.65	6.43	1.44

PENCOYD 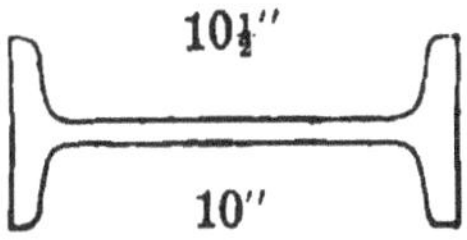 **BEAMS.**

Maximum and Minimum sections of each shape.
Greatest safe load in Net Tons evenly distributed, including beam itself.
Deflections in inches corresponding to given loads for each size of beam.
For a load in middle of beam allow one-half the tabular figures.
Deflection for latter load will be $\frac{8}{10}$ of the tabular deflection.

Chart Number.	5	5	5½	5½	6	6	Deflections for 10½″ Beams.	7	7	8	8	Deflections for 10″ Beams.
Size of Beam in Inches.	10½″	10½″	10½″	10½″	10½″	10½″		10″	10″	10″	10″	
Wt. per Yard in Lbs.	161	134	135	108	109	89		137	112	106	90	
Moment of Inertia.	265.7	241.6	219.5	195.4	180.3	162.3		194.4	173.6	161.3	148.3	
Length of Span in Feet.	Greatest Safe Load.							Greatest Safe Load				
10	23.62	21.49	19.51	17.37	16.03	13.35	.15	18.14	16.20	15.08	13.84	.16
11	21.47	19.54	17.74	15.79	14.57	13.11	.18	16.49	14.73	13.71	12.58	.19
12	19.68	17.91	16.26	14.48	13.36	12.02	.22	15.12	13.50	12.57	11.54	.23
13	18.17	16.53	15.01	13.36	12.33	11.09	.25	13.95	12.46	11.60	10.65	.27
14	16.87	15.35	13.94	12.41	11.45	10.30	.30	12.94	11.57	10.77	9.89	.31
15	15.75	14.33	13.01	11.58	10.69	9.61	.34	12.09	10.80	10.05	9.23	.36
16	14.76	13.43	12.19	10.86	10.02	9.01	.39	11.34	10.13	9.42	8.65	.41
17	13.90	12.64	11.43	10.22	9.43	8.48	.44	10.67	9.53	8.87	8.14	.46
18	13.12	11.94	10.84	9.65	8.91	8.01	.49	10.08	9.00	8.38	7.69	.52
19	12.43	11.31	10.27	9.14	8.44	7.59	.55	9.55	8.53	7.94	7.29	.58
20	11 81	10.74	9.75	8.69	8.01	7.21	.61	9.07	8.10	7.54	6.92	.64
21	11.25	10.23	9.29	8.27	7.63	6.87	.67	8.64	7.72	7.18	6.59	.71
22	10.74	9.77	8.87	7.90	7.28	6.55	.74	8.25	7.36	6.85	6.29	.78
23	10.27	9.34	8.48	7.55	6.97	6.27	.81	7.89	7.04	6.56	6.02	.85
24	9.84	8.95	8.13	7.24	6.68	6.01	.83	7.56	6.75	6.28	5.77	.92
25	9.45	8.60	7.80	6.95	6.41	5.77	.95	7.26	6.48	6.03	5.54	1.00
26	9.09	8.27	7.51	6.68	6.16	5.54	1.03	6.98	6.23	5.80	5.32	1.08
27	8.75	7.96	7.23	6.43	5.94	5.34	1.11	6.72	6.00	5.59	5.13	1.17
28	8.43	7.67	6.97	6.20	5.72	5.15	1.19	6.48	5.79	5.39	4.94	1.26
29	8.14	7.41	6.73	5.99	5.53	4.97	1.28	6.26	5.59	5.20	4.77	1.35
30	7.87	7.17	6.51	5.79	5.34	4.80	1.37	6.05	5.40	5.03	4.61	1.44
31	7.62	6.93	6.29	5.60	5.16	4.65	1.46	5.85	5.23	4.86	4.47	1.54
32	7.38	6.72	6.09	5.43	5.01	4.50	1.57	5.67	5.06	4.71	4.33	1.64
33	7.19	6.51	5.91	5.26	4.86	4.37	1.68	5.50	4.91	4.57	4.19	1.75

PENCOYD BEAMS.

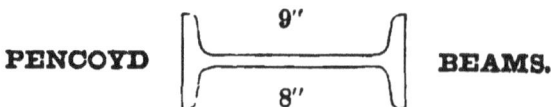

Maximum and Minimum sections of each shape.
Greatest safe load in Net Tons evenly distributed, including beam itself.
Deflection in inches corresponding to given loads for each size of beam.
For load in middle allow one-half the tabular figures.
Deflection for latter load will be $\frac{8}{10}$ of the tabular deflection.

Chart Number.	9	9	10	10	Deflections for 9" Beams.	11	11	12	12	Deflections for 8" Beams.
Size of Beam in Inches.	9"	9"	9"	9"		8"	8"	8"	8"	
Wt. per Yd. in Lbs.	122	90	88	70		109	81	75	65	
Moment of Inertia.	143.7	118.8	106.8	94.4		98.6	83.9	74.5	69.2	
	Greatest Safe Load.					Greatest Safe Load.				
6	24.27	16.53	18.42	9.94	.06	19.22	15.49	14.48	10.46	.07
7	20.80	16.53	15.79	9.94	.08	16.47	13.99	12.41	10.46	.10
8	18.20	15.40	13.81	9.94	.11	14.41	12.24	10.86	10.08	.13
9	16.18	13.69	12.27	9.94	.14	12.81	10.88	9.66	8.96	.16
10	14.56	12.32	11.05	9.79	.18	11.53	9.79	8.69	8.07	.20
11	13.24	11.20	10.04	8.90	.22	10.48	8.90	7.90	7.33	.24
12	12.13	10.26	9.21	8.16	.26	9.61	8.16	7.24	6.72	.29
13	11.20	9.48	8.50	7.53	.30	8.87	7.53	6.68	6.21	.34
14	10.40	8.80	7.89	7.00	.35	8.24	6.99	6.21	5.76	.39
15	9.71	8.21	7.37	6.53	.40	7.69	6.53	5.79	5.38	.45
16	9.10	7.70	6.91	6.12	.46	7.21	6.12	5.43	5.04	.51
17	8.56	7.25	6.50	5.76	.52	6.78	5.76	5.12	4.74	.58
18	8.09	6.84	6.14	5.44	.58	6.41	5.44	4.83	4.48	.65
19	7.66	6.48	5.82	5.15	.64	6.07	5.15	4.57	4.24	.72
20	7.28	6.16	5.52	4.90	.71	5.76	4.89	4.34	4.03	.80
21	6.93	5.86	5.25	4.66	.78	5.49	4.66	4.14	3.84	.88
22	6.62	5.60	5.02	4.45	.86	5.24	4.45	3.95	3.67	.97
23	6.33	5.35	4.80	4.25	.94	5.01	4.25	3.78	3.50	1.06
24	6.07	5.13	4.61	4.08	1.02	4.80	4.08	3.62	3.36	1.16
25	5.82	4.93	4.42	3.92	1.11	4.61	3.91	3.48	3.22	1.26
26	5.60	4.74	4.25	3.77	1.21	4.43	3.77	3.34	3.10	1.36
27	5.39	4.56	4.09	3.63	1.30	4.27	3.62	3.22	2.98	1.46
28	5.20	4.40	3.95	3.50	1.40	4.12	3.50	3.10	2.88	1.57
29	5.02	4.25	3.81	3.38	1.50	3.98	3.37	3.00	2.78	1.68

Length of Span in Feet.

PENCOYD 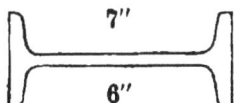 BEAMS.

7″
6″

Greatest safe load in Net Tons evenly distributed, including beam itself.
Deflections in inches corresponding to given loads for each size of beam.
For a load in middle of beam allow one-half the tabular figures.

Deflection for latter load will be $\frac{8}{10}$ of the tabular deflection.

LENGTH OF SPAN IN FEET.	CHART NUMBER. 13	13	14	14	DEFLECTIONS FOR 7″ BEAMS.	15	15	16	16	DEFLECTIONS FOR 6″ BEAMS.
SIZE OF BEAM IN INCHES.	7″	7″	7″	7″		6″	6″	6″	6″	
WT. PER YD. IN LBS.	88	75	63	51		63	55	48	40	
MOMENT OF INERTIA.	58.6	53.3	48.0	43.1		30.8	27.5	26.3	24.1	
	GREATEST SAFE LOAD.					GREATEST SAFE LOAD.				
6	12.93	11.75	10.65	6.17	.08	8.03	7.42	6.87	6.25	.10
7	11.09	10.07	9.13	6.17	.11	6.89	6.36	5.89	5.36	.13
8	9.70	8.81	7.99	6.17	.15	6.02	5.56	5.15	4.69	.17
9	8.62	7.83	7.10	6.17	.19	5.36	4.94	4.58	4.17	.22
10	7.76	7.05	6.39	5.74	.23	4.82	4.45	4.12	3.74	.27
11	7.05	6.41	5.81	5.22	.28	4.38	4.05	3.75	3.41	.32
12	6.47	5.87	5.32	4.79	.33	4.02	3.71	3.43	3.12	.38
13	5.97	5.42	4.92	4.41	.38	3.71	3.42	3.17	2.88	.45
14	5.54	5.04	4.56	4.10	.44	3.44	3.18	2.94	2.68	.52
15	5.17	4.70	4.26	3.83	.51	3.21	2.97	2.75	2.50	.60
16	4.85	4.41	3.99	3.59	.58	3.01	2.78	2.57	2.34	.69
17	4.56	4.15	3.76	3.38	.66	2.84	2.62	2.42	2.21	.78
18	4.31	3.92	3.55	3.19	.74	2.68	2.47	2.29	2.08	.87
19	4.08	3.71	3.36	3.02	.82	2.54	2.34	2.17	1.97	.97
20	3.88	3.52	3.19	2.87	.90	2.41	2.22	2.06	1.87	1.07
21	3.70	3.36	3.04	2.73	.99	2.30	2.12	1.96	1.78	1.18
22	3.53	3.20	2.90	2.61	1.09	2.19	2.02	1.87	1.70	1.29
23	3.37	3.07	2.77	2.49	1.20	2.10	1.93	1.79	1.63	1.41
24	3.23	2.94	2.66	2.39	1.32	2.01	1.85	1.72	1.56	1.54
25	3.10	2.82	2.56	2.30	1.43	1.93	1.78	1.65	1.50	1.67
26	2.98	2.71	2.46	2.21	1.55	1.85	1.71	1.58	1.44	1.81
27	2.87	2.61	2.37	2.12	1.67	1.78	1.65	1.53	1.39	1.95
28	2.77	2.52	2.28	2.05	1.80	1.72	1.59	1.47	1.34	2.10
29	2.68	2.43	2.20	1.98	1.93	1.66	1.53	1.42	1.29	2.25

PENCOYD BEAMS.

5"
4"

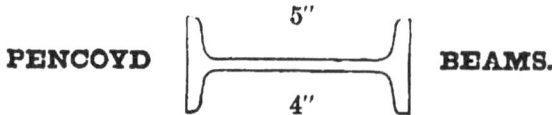

Greatest safe load in Net Tons evenly distributed, including beam itself.
Deflections in inches corresponding to given loads for each size of beam.
For a load in middle of beam allow one-half the tabular figures.

Deflection for latter load will be $\frac{8}{10}$ of the tabular deflection.

CHART NUMBER.	17	17	18	18	DEFLECTIONS FOR 5" BEAMS.	19	19	20.	20.	DEFLECTIONS FOR 4" BEAMS.
SIZE OF BEAM IN INCHES.	5"	5"	5"	5"		4"	4"	4"	4"	
WT. PER YD. IN LBS.	40	36	33	30		38	28	21.5	18.5	
MOMENT OF INERTIA.	14.7	13.7	13.1	12.5		9.0	7.7	5.5	5.1	
	GREATEST SAFE LOAD.					GREATEST SAFE LOAD.				
4	6.80	6.42	6.12	4.86	.05	5.25	4.47	3.27	3.00	.06
5	5.44	5.14	4.90	4.67	.08	4.25	3.58	2.62	2.40	.10
6	4.53	4.28	4.08	3.89	.12	3.50	2.98	2.18	2.00	.14
7	3.89	3.67	3.50	3.33	.16	3.00	2.56	1.86	1.71	.20
8	3.40	3.21	3.06	2.92	.21	2.62	2.24	1.64	1.50	.26
9	3.02	2.86	2.72	2.59	.26	2.33	1.99	1.46	1.33	.33
10	2.72	2.57	2.45	2.33	.32	2.10	1.79	1.31	1.20	.40
11	2.47	2.34	2.23	2.12	.39	1.91	1.63	1.19	1.09	.49
12	2.27	2.14	2.04	1.94	.46	1.75	1.49	1.09	1.00	.58
13	2.09	1.98	1.88	1.79	.54	1.62	1.38	1.01	.92	.68
14	1.94	1.84	1.75	1.67	.63	1.50	1.28	.94	.86	.79
15	1.81	1.71	1.63	1.55	.72	1.40	1.19	.87	.80	.91
16	1.70	1.61	1.53	1.46	.82	1.31	1.12	.82	.75	1.03
17	1.60	1.51	1.44	1.37	.93	1.23	1.05	.77	.71	1.17
18	1.51	1.43	1.36	1.30	1.04	1.17	.99	.73	.67	1.31
19	1.43	1.35	1.29	1.23	1.16	1.11	.94	.69	.63	1.46
20	1.36	1.28	1.22	1.17	1.29	1.05	.89	.65	.60	1.61
21	1.29	1.22	1.17	1.11	1.42	1.00	.85	.62	.57	1.77
22	1.24	1.17	1.11	1.06	1.56	.95	.81	.60	.54	1.93
23	1.18	1.12	1.07	1.01	1.70	.91	.78	.57	.52	2.12
24	1.13	1.07	1.02	.97	1.85	.87	.75	.55	.50	2.32
25	1.09	1.03	.98	.93	2.01	.84	.72	.52	.48	2.51
26	1.04	.99	.94	.90	2.18	.81	.69	.50	.46	2.71
27	1.01	.95	.91	.86	2.36	.78	.66	.48	.44	2.91

LENGTH OF SPAN IN FEET.

PENCOYD 3″ **BEAMS.**

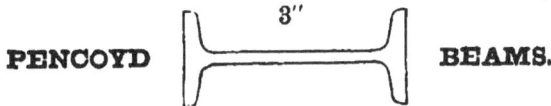

Maximum and Minimum sections of each shape.
Greatest safe load in Net Tons evenly distributed including beam itself.
Deflections in inches corresponding to given loads for each size of beam.
For a load in middle of beam allow one-half the tabular figures.

Deflection for latter load will be $\frac{8}{11}$ of the tabular deflection.

CHART NUMBER.	21	21	22	22	DEFLECTIONS FOR 3″ BEAMS.				
SIZE OF BEAM IN INCHES.	3″	3″	3″	3″					
WT. PER YD. IN LBS.	28.6	23	21.7	17					
MOMENT OF INERTIA.	4.0	3.3	3.0	2.7					
	GREATEST SAFE LOAD.								
4	2.87	2.56	2.34	2.07	.09				
5	2.30	2.05	1.87	1.66	.14				
6	1.92	1.71	1.56	1.38	.19				
7	1.64	1.46	1.34	1.18	.26				
8	1.44	1.28	1.17	1.03	.34				
9	1.28	1.14	1.04	.92	.43				
10	1.15	1.02	.94	.82	.53				
11	1.04	.93	.85	.75	.65				
12	.96	.85	.78	.69	.77				
13	.88	.79	.72	.64	.91				
14	.82	.73	.67	.59	1.05				
15	.77	.68	.62	.55	1.21				
16	.72	.64	.58	.52	1.37				
17	.68	.60	.55	.49	1.55				
18	.64	.57	.52	.46	1.74				
19	.61	.54	.49	.44	1.93				
20	.58	.51	.47	.41	2.13				
21	.55	.49	.45	.39	2.37				
22	.52	.47	.43	.38	2.62				
23	.50	.44	.41	.36	2.88				

LENGTH OF SPAN IN FEET.

PENCOYD ⌐———15″———⌐ **CHANNELS.**

12″

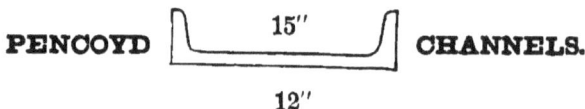

Maximum and Minimnm sections of each shape.
Greatest safe load in Net Tons evenly distributed, including beam itself.
Deflections in inches corresponding to given loads for each size of channel.
For a load in middle of beam, allow one-half the tabular figures.
Deflection for latter load will be $\frac{8}{10}$ of the tabular deflection.

CHART NUMBER.	30	30			DEFLECT'S FOR 15″ CHANNELS.	31	31	32	32	DEFLECT'S FOR 12″ CHANNELS.
SIZE OF CHANNEL IN INS.	15″	15″				12″	12″	12″	12″	
WT. PER YD. IN LBS.	204.5	148				160	88.5	101.5	60	
MOMENT OF INERTIA.	557.4	451.5				268.5	182.7	173.5	123.7	
	GREATEST SAFE LOAD.					GREATEST SAFE LOAD.				
10	34.68	28.09			.11	20.88	14.21	13.49	9.14	.13
11	31.53	25.54			.13	18.98	12.92	12.26	8.75	.16
12	28.90	23.41			.15	17.40	11.84	11.24	8.02	.19
13	26.68	21.61			.18	16.06	10.93	10.38	7.40	.22
14	24.77	20.06			.21	14.91	10.15	9.64	6.87	.26
15	23.12	19.73			.24	13.92	9.47	8.99	6.41	.30
16	21.68	17.56			.27	13.05	8.88	8.43	6.01	.34
17	20.40	16.52			.30	12.28	8.36	7.94	5.66	.38
18	19.27	15.61			.34	11.60	7.89	7.49	5.34	.43
19	18.25	14.78			.38	10.99	7.48	7.10	5.06	.48
20	17.34	14.04			.43	10.44	7.10	6.74	4.81	.53
21	16.52	13.38			.47	9.94	6.77	6.42	4.58	.58
22	15.76	12.77			.52	9.49	6.46	6.13	4.37	.64
23	15.08	12.21			.57	9.08	6.18	5.87	4.18	.70
24	14.45	11.70			.62	8.70	5.92	5.62	4.01	.77
25	13.87	11.24			.67	8.35	5.68	5.40	3.85	.83
26	13.34	10.80			.73	8.03	5.47	5.19	3.70	.90
27	12.83	10.40			.78	7.73	5.26	5.00	3.56	.97
28	12.39	10.03			.84	7.46	5.07	4.82	3.44	1.05
29	11.96	9.69			.90	7.20	4.90	4.65	3.32	1.12
30	11.56	9.36			.96	6.96	4.74	4.50	3.21	1.20
31	11.19	9.06			1.03	6.74	4.58	4.35	3.10	1.28
32	10.84	8.78			1 10	6.52	4.44	4.22	3.01	1.36
33	10.51	8.51			1.17	6.33	4.31	4.09	2.92	1.44

LENGTH OF SPAN IN FEET.

PENCOYD 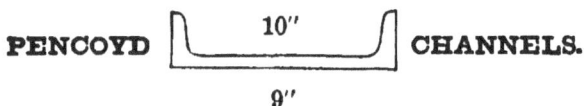 **CHANNELS.**

Maximum and Minimum sections of each shape.
Greatest safe load in Net Tons evenly distributed, including beam itself.
Deflections in inches corresponding to given loads for each size of channel.
For a load in middle of beam allow one-half the tabular figures.
Deflection for latter load will be $\tfrac{8}{10}$ of the tabular deflection.

CHART NUMBER.	34	34	35	35	DEFLECT'S FOR 10" CHANNELS.	36	36	37	37	DEFLECT'S FOR 9" CHANNELS.
SIZE OF CHANNEL IN INS.	10"	10"	10"	10"		9"	9"	9"	9"	
WT. PER YD. IN LBS.	106	60	86.5	49		93	54	61	37	
MOMENT OF INERTIA.	131.0	92.7	105.2	73.9		90.7	64.3	59.8	43.6	
	GREATEST SAFE LOAD.					GREATEST SAFE LOAD.				
LENGTH OF SPAN IN FEET.										
10	12.23	8.65	9.81	6.89	.16	9.41	6.67	6.21	4.52	.18
11	11.12	7.86	8.92	6.26	.19	8.55	6.06	5.65	4.11	.22
12	10.19	7.21	8.17	5.74	.23	7.84	5.56	5.17	3.77	.26
13	9.41	6.65	7.55	5.30	.27	7.24	5.13	4.78	3.48	.30
14	8.74	6.18	7.01	4.92	.31	6.72	4.76	4.44	3.23	.35
15	8.15	5.77	6.54	4.59	.36	6.27	4.45	4.14	3.01	.40
16	7.64	5.41	6.13	4.31	.41	5.88	4.17	3.88	2.82	.46
17	7.19	5.09	5.77	4.05	.46	5.53	3.92	3.65	2.66	.52
18	6.79	4.81	5.45	3.83	.52	5.23	3.71	3.45	2.51	.58
19	6.44	4.55	5.16	3.63	.58	4.95	3.51	3.27	2.38	.64
20	6.11	4.32	4.90	3.44	.64	4.70	3.34	3.10	2.26	.71
21	5.82	4.12	4.67	3.28	.71	4.48	3.18	2.96	2.15	.78
22	5.56	3.93	4.46	3.13	.78	4.28	3.03	2.82	2.05	.86
23	5.32	3.76	4.27	2.99	.85	4.09	2.90	2.70	1.97	.94
24	5.10	3.60	4.09	2.87	.92	3.92	2.78	2.59	1.88	1.02
25	4.89	3.46	3.92	2.76	1.00	3.76	2.67	2.48	1.81	1.11
26	4.70	3.33	3.77	2.65	1.08	3.62	2.57	2.39	1.74	1.21
27	4.53	3.20	3.63	2.55	1.17	3.49	2.47	2.30	1.67	1.30
28	4.37	3.09	3.50	2.46	1.26	3.36	2.38	2.22	1.61	1.40
29	4.22	2.98	3.38	2.38	1.35	3.24	2.30	2.14	1.56	1.50
30	4.08	2.88	3.27	2.30	1.45	3.14	2.22	2.07	1.51	1.61
31	3.94	2.79	3.16	2.22	1.55	3.03	2.15	2.00	1.46	1.72
32	3.82	2.70	3.07	2.15	1.65	2.94	2.08	1.94	1.41	1.83
33	3.71	2.62	2.97	2.09	1.76	2.85	2.02	1.88	1.37	1.95

PENCOYD ⌐___8″___⌐ CHANNELS.

7″

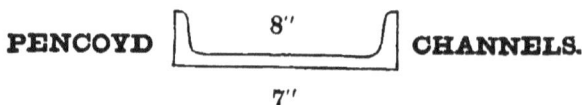

Maximum and Minimum sections of each shape.

Greatest safe load in Net Tons evenly distributed, including beam itself.

Deflection in inches corresponding to given loads for each size of channel.

For load in middle of beam allow one-half the tabular figures.

Deflection for latter load will be $\frac{8}{10}$ of the tabular deflection.

CHART NUMBER.	38	38	39	39	DEFLECTIONS FOR 8″ CHANNELS.	40	40	41	41	DEFLECTIONS FOR 7″ CHANNELS.
SIZE OF CHANNEL IN INS.	8″	8″	8″	8″		7″	7″	7″	7″	
WT. PER YD. IN LBS.	80.5	43	54	30		73	41	49	26	
MOMENT OF INERTIA.	60.0	40.0	41.0	28.2		42.6	29.5	27.9	18.5	
LENGTH OF SPAN IN FEET.	GREATEST SAFE LOAD.					GREATEST SAFE LOAD.				
6	11.67	7.77	7.93	4.79	.07	9.45	6.55	6.18	3.42	.08
7	10.00	6.66	6.83	4.70	.10	8.10	5.61	5.30	3.42	.11
8	8.75	5.83	5.97	4.11	.13	7.09	4.91	4.64	3.07	.15
9	7.78	5.18	5.31	3.66	.16	6.30	4.37	4.12	2.73	.19
10	7.00	4.66	4.78	3.29	.20	5.67	3.93	3.71	2.46	.23
11	6.36	4.24	4.35	2.99	.24	5.15	3.57	3.37	2.24	.28
12	5.83	3.88	3.98	2.74	.29	4.72	3.27	3.09	2.05	.33
13	5.38	3.58	3.67	2.53	.34	4.36	3.02	2.85	1.89	.38
14	5.00	3.33	3.41	2.35	.39	4.05	2.81	2.65	1.76	.45
15	4.67	3.11	3.19	2.19	.45	3.78	2.62	2.47	1.64	.52
16	4.37	2.91	2.99	2.06	.51	3.54	2.46	2.32	1.54	.59
17	4.12	2.74	2.81	1.94	.58	3.34	2.31	2.18	1.45	.67
18	3.89	2.59	2.66	1.83	.65	3.15	2.18	2.06	1.37	.75
19	3.68	2.45	2.52	1.73	.72	2.98	2.07	1.95	1.29	.83
20	3.50	2.33	2.39	1.64	.80	2.83	1.96	1.85	1.23	.92
21	3.33	2.22	2.28	1.57	.88	2.70	1.87	1.77	1.17	1.01
22	3.18	2.12	2.17	1.50	.97	2.58	1.79	1.69	1.12	1.11
23	3.04	2.03	2.08	1.43	1.06	2.47	1.71	1.61	1.07	1.22
24	2.92	1.94	1.99	1.37	1.16	2.36	1.64	1.55	1.02	1.34
25	2.80	1.86	1.91	1.32	1.26	2.27	1.57	1.48	.98	1.45
26	2.60	1.79	1.84	1.26	1.36	2.18	1.51	1.43	.95	1.57
27	2.59	1.73	1.77	1.22	1.46	2.10	1.46	1.37	.91	1.69
28	2.50	1.66	1.71	1.17	1.57	2.02	1.40	1.32	.88	1.82
29	2.41	1.61	1.65	1.13	1.68	1.95	1.35	1.28	.85	1.95

PENCOYD 5" and 6" / 3" and 4' CHANNELS.

Maximum and Minimum sections of each shape.
Greatest safe load in Net Tons evenly distributed including beam itself.
Deflections in inches corresponding to given loads for each size of channel.
For a load in middle of beam allow one-half the tabular figures.
Deflection for latter load will be $\frac{8}{10}$ of the tabular deflection. ·

CHART NUMBER.	DEFLECT'S FOR 6" CHANNELS.	42	44	45	46	DEFLECT'S FOR 5" CHANNELS.	DEFLECT'S FOR 4" CHANNELS.	47	48	49	49	DEFLECT'S FOR 3" CHANNELS.
SIZE OF CHANNEL IN INS.		6"	6"	5"	5"			4"	4"	3"	3"	
WT. PER YD. IN LBS.		33.0	23	27	18.8			21.5	17.5	18.9	15.2	
MOM. OF INERTIA.		18.4	11.7	10.3	6.7			5.2	4.1	2.3	2.0	
		GREATEST SAFE LOAD.						GREATEST SAFE LOAD.				
4	.02	6.50	5.24	5.92	4.13	.02	.03	4.03	3.20	2.40	2.10	.05
	.04	6.50	4.54	4.80	3.10	.05	.06	3.02	2.40	1.80	1.57	.09
5	.07	5.70	3.63	3.84	2.48	.08	.10	2.42	1.92	1.44	1.26	.14
6	.10	4.75	3.02	3.20	2.07	.12	.14	2.02	1.60	1.20	1.05	.19
7	.13	4.07	2.59	2.74	1.77	.16	.20	1.73	1.37	1.03	.90	.26
8	.17	3.56	2.26	2.40	1.55	.21	.26	1.51	1.20	.90	.79	.34
9	.22	3.17	2.01	2.13	1.38	.26	.33	1.34	1.07	.80	.70	.43
10	.27	2.85	1.81	1.92	1.24	.32	.40	1.21	.96	.72	.63	.53
11	.32	2.59	1.64	1.74	1.13	.39	.49	1.10	.87	.65	.57	.65
12	.38	2.37	1.51	1.60	1.03	.46	.58	1.01	.80	.60	.52	.77
13	.45	2.19	1.39	1.48	.95	.54	.68	.93	.74	.55	.48	.91
14	.52	2.04	1.29	1.37	.89	.63	.79	.86	.69	.51	.45	1.05
15	.60	1.90	1.21	1.28	.83	.72	.91	.81	.64	.48	.42	1.21
16	.69	1.78	1.13	1.20	.77	.82	1.03	.76	.60	.45	.39	1.37
17	.78	1.68	1.06	1.13	.73	.93	1.17	.71	.56	.42	.37	1.55
18	.87	1.58	1.01	1.07	.69	1.04	1.31	.67	.53	.40	.35	1.74
19	.97	1.50	.95	1.01	.65	1.16	1.46	.64	.51	.38	.33	1.93
20	1.07	1.42	.90	.96	.62	1.29	1.61	.60	.48	.36	.31	2.13
21	1.18	1.36	.86	.91	.59	1.42	1.77	.58	.46	.34	.30	2.37
22	1.29	1.30	.82	.87	.56	1.56	1.93	.55	.44	.33	.29	2.62
23	1.41	1.24	.79	.83	.54	1.70	2.12	.53	.42	.31	.27	2.88
24	1.54	1.19	.75	.80	.52	1.85	2.32	.50	.40	.30	.26	3.11
25	1.67	1.14	.72	.77	.50	2.01	2.51	.48	.38	.29	.25	3.34
26	1.81	1.10	.70	.74	.48	2.18	2.71	.47	.37	.28	.24	3.50

LENGTH OF SPAN IN FEET.

4

PENCOYD BEAMS.

12" and 11"
10" and 9"

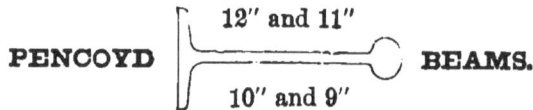

Maximum and minimum sections of each shape.
Greatest safe load in Net Tons evenly distributed, including beam itself.
Deflections in inches corresponding to given loads for each size of beam.
For a load-in middle of beam allow one-half the tabular figures.

Deflection for latter load will be $\frac{5}{10}$ of the tabular deflection.

LENGTH OF SPAN IN FEET.	DEFLECTIONS FOR 12" BEAMS.	60 12"	60 12"	61 11"	61 11"	DEFLECTIONS FOR 11" BEAMS.	DEFLECTIONS FOR 10" BEAMS.	62 10"	62 10"	63 9"	63 9"	DEFLECTIONS FOR 9" BEAMS.
SIZE OF BEAM IN INCHES.		12"	12"	11"	11"			10"	10"	9"	9"	
WT. PER YD. IN LBS.		138	104	118	91			105	80	94	72	
MOM. OF INERTIA.		264.9	222.0	193.1	164.1			140.4	118.2	99.5	84.8	
		GREATEST SAFE LOAD.						GREATEST SAFE LOAD.				
10	.13	20.59	17.26	16.41	13.95	.15	.16	13.11	11.03	10.32	8.79	.18
11	.16	18.72	15.69	14.92	12.68	.18	.19	11.92	10.03	9.38	7.99	.22
12	.19	17.16	14.38	13.67	11.62	.21	.23	10.92	9.19	8.60	7.32	.26
13	.22	15.84	13.28	12.62	10.73	.25	.27	10.08	8.48	7.94	6.76	.30
14	.26	14.71	12.33	11.72	9.96	.29	.31	9.36	7.88	7.37	6.28	.35
15	.30	13.73	11.51	10.94	9.30	.34	.36	8.74	7.35	6.88	5.86	.40
16	.34	12.87	10.79	10.26	8.72	.39	.41	8.19	6.89	6.45	5.49	.47
17	.39	12.11	10.15	9.65	8.21	.44	.46	7.71	6.49	6.07	5.17	.53
18	.44	11.44	9.59	9.12	7.75	.49	.52	7.28	6.18	5.73	4.88	.59
19	.49	10.84	9.08	8.64	7.34	.54	.58	6.90	5.81	5.43	4.63	.65
20	.54	10.29	8.63	8.20	6.97	.59	.64	6.55	5.51	5.16	4.39	.72
21	.59	9.80	8.22	7.81	6.64	.65	.71	6.24	5.25	4.91	4.19	.79
22	.65	9.36	7.85	7.46	6.34	.71	.78	5.96	5.01	4.69	4.00	.87
23	.71	8.95	7.50	7.15	6.07	.77	.86	5.70	4.80	4.49	3.82	.95
24	.78	8.58	7.19	6.84	5.81	.84	.93	5.46	4.60	4.30	3.66	1.04
25	.84	8.24	6.90	6.56	5.58	.91	1.01	5.24	4.41	4.13	3.52	1.13
26	.92	7.92	6.64	6.31	5.37	.99	1.09	5.04	4.24	3.97	3.38	1.23
27	.93	7.63	6.39	6.08	5.17	1.07	1.18	4.86	4.09	3.82	3.26	1.32
28	1.07	7.35	6.16	5.86	4.96	1.15	1.27	4.68	3.94	3.69	3.14	1.42
29	1.14	7.10	5.95	5.66	4.81	1.23	1.36	4.52	3.80	3.56	3.03	1.52
30	1.22	6.86	5.75	5.47	4.65	1.32	1.45	4.37	3.67	3.44	2.93	1.65
31	1.30	6.64	5.57	5.29	4.50	1.41	1.55	4.23	3.56	3.33	2.84	1.77
32	1.33	6.43	5.39	5.13	4.36	1.50	1.65	4.10	3.45	3.22	2.75	1.90
33	1.46	6.24	5.23	4.97	4.23	1.60	1.76	3.97	3.34	3.18	2.66	2.03

PENCOYD [8" and 7" / 6" and 5"] BEAMS.

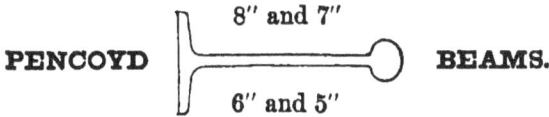

Maximum and minimum sections of each shape.
Greatest safe load in Net Tons evenly distributed, including beam itself.
Deflections in inches corresponding to given loads for each size of beam.
For a load in middle of beam allow one-half the tabular figures.
Deflection for latter load will be $\frac{5}{10}$ of the tabular deflection.

	Defl. for 8" Beams		Chart No. 64	64	65	65	Defl. for 7" Beams	Defl. for 6" Beams	66	66	67	67	Defl. for 5" Beams
Size of beam in Ins.			8"	8"	7"	7"			6"	6"	5"	5"	
Wt. per Yd. in Lbs.			84	61	72	52			57	42	46	34	
Mom. of Inertia			70.5	57.7	42.6	34.4			26.5	22.0	14.5	12.0	
Length of span in feet			Greatest Safe Load						Greatest Safe Load				
6	.07		19.53	11.22	9.43	7.63	.08	.10	6.87	5.70	4.53	3.73	.12
7	.10		16.74	9.61	8.09	6.54	.11	.13	5.89	4.89	3.89	3.20	.16
8	.13		14.65	8.41	7.07	5.73	.15	.17	5.15	4.27	3.40	2.80	.21
9	.16		13.02	7.48	6.29	5.09	.19	.22	4.58	3.80	3.02	2.49	.26
10	.20		11.72	6.73	5.66	4.58	.23	.27	4.12	3.42	2.72	2.24	.32
11	.24		10.65	6.12	5.15	4.16	.28	.32	3.75	3.11	2.47	2.04	.39
12	.29		9.77	5.61	4.72	3.82	.33	.38	3.43	2.85	2.27	1.87	.46
13	.34		9.01	5.18	4.35	3.52	.38	.45	3.17	2.63	2.09	1.72	.54
14	.39		8.37	4.81	4.04	3.27	.45	.52	2.94	2.44	1.94	1.60	.63
15	.45		7.81	4.49	3.77	3.05	.52	.59	2.75	2.28	1.81	1.49	.72
16	.51		7.32	4.21	3.54	2.86	.59	.69	2.57	2.14	1.70	1.40	82
17	.58		6.89	3.96	3.33	2.69	.67	.78	2.42	2.01	1.60	1.32	.93
18	.65		6.51	3.74	3.14	2.54	.75	.87	2.29	1.90	1.51	1.24	1.04
19	.72		6.17	3.54	2.98	2.41	.83	.97	2.17	1.80	1.43	1.18	1.16
20	.80		5.86	3.36	2.83	2.29	.92	1.07	2.06	1.71	1.36	1.12	1.29
21	.88		5.58	3.20	2.69	2.18	1.01	1.18	1.96	1.63	1.30	1.07	1.42
22	.97		5.33	3.06	2.57	2.08	1.11	1.29	1.87	1.55	1.24	1.02	1.56
23	1.06		5.10	2.93	2.46	1.99	1.22	1.41	1.79	1.49	1.18	.97	1.70
24	1.16		4.88	2.80	2.36	1.91	1.34	1.54	1.72	1.42	1.13	.93	1.85
25	1.26		4.69	2.69	2.26	1.83	1.45	1.67	1.65	1.37	1.09	.90	2.01
26	1.36		4.51	2.59	2.18	1.76	1.57	1.81	1.58	1.32	1.05	.86	2.18
27	1.46		4.34	2.49	2.10	1.70	1.69	1.95	1.53	1.27	1.01	.83	2.36
28	1.57		4.19	2.40	2.02	1.64	1.82	2.10	1.47	1.22	.97	.80	2.54
29	1.68		4.04	2.32	1.95	1.58	1.95	2.25	1.42	1.18	.94	.77	2.73

IRON FLOOR BEAMS.

When I beams are used as floor joists or girders, the spacing and proper size of beams depends on the amount and character of the loads, as well as the distance to be spanned. Not only the positive strength, but the elasticity or amount of deflection permissible must be considered.

A heavy load per unit of area may not require as strong a floor as that necessary for a lighter one, if the latter be liable to sudden application, especially if accompanied with impact, while the normal state of the heavier load is quiescence, or slow and even change. It would require a special treatise to describe the subject, and those lacking experience are referred to the published literature which is now very ample and complete. It has been demonstrated that the greatest mass of men that can be packed on any floor will not exceed in weight 80 lbs. per square foot. The weight of the iron beams will depend on the span, for which see a general rule farther on. If brick arches are laid between the beams, the weight of a 4″ course of brick, including the concrete filling, will be about 50 lbs. per square foot.

Within the limits of length of span in which rolled I beams can be used, it may be assumed that a floor is safe to sustain the greatest possible load of men, when the following loading does not exhibit a greater bending stress on the beam than that denoted in the tables, under the head of "Greatest Safe Load Distributed," pages 40-51.

I Beam joists with wooden floor   = 100 lbs. per square foot.
Wooden floor and plastered ceilings = 110 "   "   "   "
4″ brick arches and concrete filling = 150 "   "   "   "

These figures represent the total weight of floor itself and the imposed load.

When the floor beams are subject to the action of moving loads, it is necessary to make allowance for a greater nominal weight than actually may occur, especially if the span is long in proportion to the depth of the beam. If the beams are too light, the resulting tremor and vibration will be a source of discomfort to the user, if not of weakness to the structure. The same results are obtained by assuming either a higher nominal load per unit of area than actually can occur, or adopting a higher factor of safety, than given in our tables, for the actual loads. Floors proportioned as follows for given purposes will be found satisfactory. The weight of the material may be included in the figures.

CHARACTER OF FLOOR.	LOAD PER SQ. FT.
Very lightest floors, plank covering............	100 lbs.
Very lightest floors, brick arches..............	150 "
Light warehouse floors........................	200 "
Halls of audience......	200 "
Warehouses in which heavy pieces are moved..	250 "
Shop floors for light machinery...............	250 "
Shop floors for heavy machinery..............	300 to 500 lbs.

GENERAL RULE FOR THE WEIGHT OF IRON IN FLOOR BEAMS.

When the standard section of any size of beam is used, the weight of iron obtained by the following rule will be found to approximate closely to the actual amount required: "Square of span in feet divided by 5 times the depth of the beam in inches, equals the pounds of iron in the beams per square foot of floor" $\left(\dfrac{\text{span}^2}{5 \times \text{depth}} = \text{lbs.} \right)$

This is for a load of 150 lbs. per square foot, and the beams strained up to the maximum safe limit as given in the tables.

With the same space the weight of the beams will vary directly as the load varies, consequently the weight of iron for any other required loading per square foot can be obtained by proportion from above rule. *Example.*—A floor of 20 feet span is subject to a load of 150 lbs. per square foot. The weight of the iron beams will be $\dfrac{20^2}{5 \times 15} = 5.33$ lbs. per square foot of floor, if 15″

Beams are used, or if 12″ Beams are used $\dfrac{20^2}{5 \times 12} = 6.66$ lbs. per

square foot. To these figures add the weight of ends built into the wall, which should be from 6″ to 12″ at each end, according to the span, etc. If the load to be sustained is 250 lbs. per sq. foot, on 15″ I beams the necessary weight becomes as 150 : 250 :: 5.33 lbs. : 8.88 lbs. per square foot.

This rule applies only to the minimum section of any I beam. If the section is increased, the weight of iron required will also increase. By the above it will be observed that the deeper the beam used the less the amount of iron required, and such is the case as a general rule. But for short spans the use of the deepest beams might require too wide a spacing to suit the covering of the floor. Then the best economy requires the adoption of a shallower and lighter beam. For brick arches for fire-proof floors it is usual to limit the rise or spring from 3 to 6 inches, in order to build in and conceal the tie rods, which should not be much if any above the center of the beam. For such flat arches the spacing of the beams should not exceed 6 feet, and if a single 4″ course of brick is used, it is safest not to exceed 5 feet separation. Of course for arches of more rise and for other special purposes than indicated above, no such limitation is necessary.

SPACING OF FLOOR BEAMS.

The following rule gives the greatest distance apart that floor-beams can be placed to support safely any given load per square foot. Multiply the length of span in feet by the load in lbs. per square foot. Find in the table, page 40, the safe load in lbs. for a beam of the size and length desirable to use. Divide this safe load by the product first found, and the quotient is the greatest distance in feet that the beams ought to be placed, center to center. Or Distance $= \dfrac{\text{Safe Load}}{w\,L^2}$. $w =$ lbs. per square foot. $L =$ length of span in feet.

Example.—A floor of 20 feet span with its full load will weigh 150 lbs. per square foot. Different sizes of beams may be safely spaced as far apart as follows: For 15″ — 145 lb. I Beams $\dfrac{32430}{20 \times 150} = 10.8$ feet center to center. For 12″ 120 lb. I beams $\dfrac{21220}{20 \times 150} = 7.07$ feet, etc., etc.

The tables on pages 56-62 show the greatest distance apart, center to center, that beams should be placed for a loading (including the weight of the floor itself) of 100, 150, 200, or 250 lbs. per square foot.

The deflections of the beams which are given in the tables will be uniform for beams of the given spans so long as the spacing is proportioned according to the table.

In the case of plastered ceilings or other circumstances where undue deflection might be injurious, it is considered good practice to limit the deflection to about $\frac{1}{360}$ of the span. When the deflections exceed this amount, the corresponding loads in the table are printed in small figures. When the deflection is below this amount, the figures for the loads are in larger print. The proper spacing of beams for any load is inversely proportioned to the loads. Consequently the proper distance apart for beams for any load per square foot can be easily obtained directly from the table as well as by the rule previously given.

Rule.—Multiply the distance given in the table by 150 and divide by the number of lbs. per square foot required to be sustained. The quotient will be the greatest distance apart for the beams.

Example.—What is the greatest distance apart 8″ 65 lbs. I beams can be placed to support safely a load of 220 lbs. per square foot, the beams having a clear span of 18 feet ? By the table the spacing for 150 lbs. per foot is 3.3 feet $\dfrac{3.3 \times 150}{220} = 2.25$ feet, the distance required.

PENCOYD  **DECK BEAMS.**

Greatest distance between floor beams so that the bending stress on the beam will not exceed its maximum safe load.

Chart Number.	Size of beam in inches.	Weight per yard, lbs.	Load per sq. ft. of Floor, lbs.	Length of span in feet. Distance between centres of beams in feet.					
				10	12	14	16	18	20
60	12	104	100			17.6	13.5	10.7	8.6
			150			11.7	9.0	7.1	5.8
			200			8.8	6.7	5.3	4.3
			250			7.0	5.4	4.3	3.5
	Deflection	in Inches.		......	......	.28	.34	.44	.54
61	11	91	100		19.3	14.2	10.9	8.6	7.0
			150		12.9	9.5	7.2	5.7	4.6
			200		9.7	7.1	5.4	4.3	3.5
			250		7.7	5.7	4.3	3.4	2.8
	Deflection	in Inches.		......	.21	.29	.37	.46	.58
62	10	80	100	22.1	15.3	11.3	8.6	6.8	5.5
			150	14.7	10.2	7.5	5.7	4.5	3.7
			200	11.0	7.7	5.6	4.3	3.4	2.8
			250	8.8	6.1	4.5	3.4	2.7	2.2
	Deflection	in Inches.		.16	.23	.32	.41	.52	.64
63	9	72	100	17.6	12.2	9.0	6.9	5.4	4·4
			150	11.7	8.1	6.0	4.6	3.6	2·9
			200	8.8	6.1	4.5	3.4	2.7	2·2
			250	7.0	4.9	3.6	2.7	2.2	1·8
	Deflection	in Inches.		.18	.26	.35	.46	.58	·71
64	8	61	100	13.4	9.3	6.9	5.2	4·1	3·4
			150	9.0	6.2	4.6	3.5	2·8	2·2
			200	6.7	4.7	3.4	2.6	2·1	1·7
			250	5.4	3.7	2.7	2.1	1·7	1·3
	Deflection	in Inches.		.20	.29	.39	.51	·65	·80
65	7	52	100	9.2	6.4	4.7	3·6	2·8	2·3
			150	6.1	4.2	3.1	2·4	1·9	1·5
			200	4.6	3.2	2.3	1·8	1·4	1·1
			250	3.7	2.5	1.9	1·4	1·1	·9
	Deflection	in Inches.		.23	.33	.45	·59	·75	·92
66	6	42	100	6.8	4.7	3·5	2·7	2·1	1·7
			150	4.5	3.2	2·8	1·5	1·4	1·1
			200	3.4	2.4	1·7	1·3	1·1	·9
			250	2.7	1.9	1·4	1·1	·8	·7
	Deflection	in Inches.		.27	.39	·52	·69	·87	1·07
67	5	34	100	4.5	3·1	2·3	1·8		
			150	3.0	2·1	1·5	1·2		
			200	2.2	1·6	1·1	·9		
			250	1.8	1·2	·9	·7		
	Deflection	in Inches.		.32	·46	·63	·82	......	....

PENCOYD   DECK BEAMS.

Figures in small type denote that the beams so placed will deflect more than $\frac{1}{16}$ of an inch for each foot of span.

22	24	26	28	30	32	LOAD PER SQ. FT. OF FLOOR, LBS.	WEIGHT PER YARD, LBS.	SIZE OF BEAM IN INCHES.	CHART NUMBER.
LENGTH OF SPAN IN FEET.									
DISTANCE BETWEEN CENTRES OF BEAMS IN FEET.									
7.1	6.0	5·1	4·4	3·8	3·4	100			
4.8	4.0	3·4	2·9	2·6	2·2	150			
3.6	3.0	2·6	2·2	1·9	1·7	200	104	12	60
2.9	2.4	2·0	1·8	1·5	1·3	250			
.65	.78	·91	1·06	1·21	1·38	......	Deflection	in Inches.	
5.8	4·8	4·1	3·6	3·1	2·7	100			
3.8	3·2	2·7	2·4	2·1	1·8	150			
2.9	2·4	2·1	1·8	1·5	1·4	200	91	11	61
2.3	1·9	1·6	1·4	1·2	1·1	250			
.71	·84	·99	1·16	1·32	1·50	......	Deflection	in Inches.	
4·6	3·8	3·3	2·8	2·5		100			
3·0	2·6	2·2	1·9	1·6		150			
2·3	1·9	1·6	1·4	1·2		200	80	10	62
1·8	1·5	1·3	1·1	1·0		250			
·78	·93	1·09	1·26	1·45	......	......	Deflection	in Inches.	
3·6	3·1	2·6	2·2			100			
2·4	2·0	1·7	1·5			150			
1·6	1·5	1·2	1·1			200	72	9	63
1·5	1·2	1·0	·9			250			
·86	1·03	1·21	1·40	......	......	.. ...	Deflection	in Inches.	
2·8	2·3	2·0				100			
1·9	1·6	1·3				150			
1·4	1·2	1·0				200	61	8	64
1·1	·9	·8				250			
·97	1·16	1·36	......	......	......	......	Deflection	in Inches.	
1·9						100			
1·3						150			
·9						200	52	7	65
·8						250			
1·30	......	......	......	......	......	......	Deflection	in Inches.	
						100			
						150			
						200	42	6	66
						250			
.....	......	......	......	......	......	......	Deflection	in Inches.	
						100			
						150			
						200	34	5	67
						250			
.....	.....	......	... ...	......	......	......	Deflection	in Inches.	

PENCOYD  BEAMS.

Greatest distances between centres of floor beams, so that the bending stress on the beam will not exceed its maximum safe load.

Chart Number	Size of Beam in Inches.	Weight per Yard, Lbs.	Load per sq. ft. of Floor, Lbs.	Length of Span in Feet.					
				10	12	14	16	18	20
				Distance between centres of Beams in Feet.					
1	15	200	100				33.1	26.2	21.2
			150				22.1	17.5	14.1
			200				16.6	13.1	10.6
			250				13.3	10.5	8.5
	Deflection	in Inches.	......	......	.....	......	.27	.34	.42
2	15	145	100				25.3	20.0	16.2
			150				16.9	13.3	10.8
			200				12.7	10.0	8.1
			250				10.1	8.0	6.5
	Deflection	in Inches.	......	......	.....	.....	.27	.34	.42
3	12	168	100			29.5	22.6	17.9	14.5
			150			19.7	15.1	11.9	9.6
			200			14.8	11.3	8.9	7.2
			250			11.8	9.0	7.1	5.8
	Deflection	in Inches.	......	......	......	.26	.34	.43	.53
4	12	120	100			21.7	16.6	13.1	10.6
			150			14.4	11.1	8.7	7.1
			200			10.8	8.3	6.5	5.3
			250			8.7	6.6	5.2	4.2
	Deflection	in Inches.	......	......	......	.26	.34	.43	.53
5	10½	134	100		29.8	21.9	16.8	13.3	10.7
			150		19.9	14.6	11.2	8.9	7.1
			200		14.9	11.0	8.4	6.6	5.4
			250		11.9	8.8	6.7	5.3	4.3
	Deflection	in Inches.	......	.22	.30	.39	.49	.61	
5½	10½	108	100		24.1	17.7	13.6	10.7	8.7
			150		16.1	11.8	9.1	7.1	5.8
			200		12.1	8.9	6.8	5.4	4.3
			250		9.7	7.1	5.4	4.3	3.4
	Deflection	in Inches.	......	.22	.30	.39	.49	.61	
6	10½	89	100		20.0	14.7	11.3	8.9	7.2
			150		13.3	9.8	7.5	5.9	4.8
			200		10.0	7.4	5.6	4.4	3.6
			250		8.0	5.9	4.5	3.5	2.9
	Deflection	in Inches	......	.22	.30	.39	.49	.61	

PENCOYD  BEAMS.

Figures in small type denote that the beams so placed will deflect more than $\frac{1}{30}$ of an inch for each foot of span.

LENGTH OF SPAN IN FEET.						LOAD PER SQ. FT. OF FLOOR, LBS.	WEIGHT PER YARD, LBS.	SIZE OF BEAM IN INCHES.	CHART NUMBER.
22	24	26	29	30	32				
DISTANCE BETWEEN CENTRES OF BEAMS IN FEET.									
17.5	14.7	12.5	10.8	9.4	8·3	100			
11.7	9.8	8.4	7.2	6.3	5·5	150			
8.8	7.4	6.3	5.4	4.7	4·1	200	200	15	1
7.0	5.9	5.0	4.3	3.8	3·3	250			
.51	.61	.72	.83	.95	1·09	......	Deflection	in Inches.	
13.4	11.3	9.6	8.3	7.2	6·3	100			
8.9	7.5	6.4	5.5	4.8	4·2	150			
6.7	5.6	4.8	4.1	3.6	3·2	200	145	15	2
5.4	4.5	3.9	3.4	2.9	2·5	250			
.51	.61	.72	.83	.95	1·09	......	Deflection	in Inches.	
12.0	10.0	8·6	7·4	6·4	5·7	100			
8.0	6.7	5·7	4·9	4·3	3·8	150			
6.0	5.0	4·3	3·7	3·2	2·8	200	168	12	3
4.8	4.0	3·4	3·0	2·6	2·3	250			
.64	.77	·90	1·05	1·20	1·36	......	Deflection	in Inches.	
8.8	7.4	6·3	5·4	4.7	4·1	100			
5.8	4.9	4·2	3·6	3·1	2·8	150			
4.4	3.7	3·1	2·7	2·4	2·1	2 0	120	12	4
3.5	2.9	2·5	2·2	1·9	1·7	250			
.64	.77	·30	1·05	1·20	1·36	......	Deflection	in Inches.	
8·9	7·5	6·4	5·6	4·8	4·2	100			
5·9	5·0	4·3	3·7	3·2	2·8	150			
4·4	3·7	3·2	2·7	2·4	2·1	200	134	10½	5
3·5	3·0	2·6	2·2	1·9	1·7	250			
·74	·88	1·03	1·19	1·37	1·67	......	Deflection	in Inches.	
7·2	6·0	5·1	4·4	3·9	3·4	100			
4·8	4·0	3·4	2·9	2·6	2·3	150			
3·6	3·0	2·6	2·2	1·9	1·7	200	108	10½	5½
2·9	2·4	2·1	1·8	1·6	1·4	250			
·74	·88	1·03	1·19	1·37	1·67	......	Deflection	in Inches.	
6·0	5·0	4·3	3·7	3·2	2·8	100			
4·0	3·3	2·9	2·5	2·1	1·9	150			
3·0	2·5	2·1	1·8	1·6	1·4	200	89	10½	6
2·4	2·0	1·7	1·4	1·3	1·1	250			
·74	·88	1·03	1·19	1·37	1·67	......	Deflection	in Inches.	

PENCOYD  BEAMS.

Greatest distances between centres of floor beams, so that the bending stress on the beam will not exceed its maximum safe load.

Chart Number.	Size of Beam in Inches.	Weight per Yard, LBS.	Load per sq. ft. of Floor, LBS.	Length of Span in Feet.					
				10	12	14	16	18	20
				Distance between centres of Beams in Feet.					
7	10	112	100	32.4	29.5	16.5	12.7	10.0	8.1
			150	21.6	15.0	11.0	8.4	6.7	5.4
			200	16.2	11.3	8.3	6.3	5.0	4.1
			250	13.0	9.0	6.6	5.1	4.0	3.2
	Deflection	in Inches.	......	.16	.23	.31	.41	.52	.64
8	10	90	100	27.7	19.2	14.1	10.8	8.5	6.9
			150	18.4	12.8	9.4	7.2	5.7	4.6
			200	13.8	9.6	7.1	5.4	4.3	3.5
			250	11.1	7.7	5.6	4.3	3.4	2.8
	Deflection	in Inches.	......	.16	.23	.31	.41	.52	.64
9	9	90	100	24.6	17.1	12.6	9.6	7.6	6.2
			150	16.4	11.4	8.4	6.4	5.1	4.1
			200	12.3	8.6	6.3	4.8	3.8	3.1
			250	9.9	6.8	5.0	3.8	3.0	2.5
	Deflection	in Inches.	......	.18	.26	.35	.46	.58	.71
10	9	70	100	19.6	13.6	10.0	7.7	6.1	4.9
			150	13.1	9.1	6.7	5.1	4.0	3.3
			200	9.8	6.8	5.0	3.8	3.0	2.4
			250	7.8	5.4	4.0	3.1	2.4	2.0
	Deflection	in Inches.	......	.18	.26	.35	.46	.58	.71
11	8	81	100	19.6	13.6	10.0	7.7	6.1	4.9
			150	13.1	9.1	6.7	5.1	4.0	3.3
			200	9.8	6.8	5.0	3.8	3.0	2.4
			250	7.8	5.4	4.0	3.1	2.4	2.0
	Deflection	in Inches.	......	.20	.29	.39	.51	.65	.80
12	8	65	100	16.1	11.2	8.2	6.3	5.0	4.0
			150	10.7	7.5	5.5	4.2	3.3	2.7
			200	8.1	5.6	4.1	3.2	2.5	2.0
			250	6.5	4.5	3.3	2.5	2.0	1.6
	Deflection	in Inches.	......	.20	.29	.39	.51	.65	.80
13	7	65	100	13.3	9.2	6.8	5.2	4.1	3.3
			150	8.8	6.1	4.5	3.5	2.7	2.2
			200	6.6	4.6	3.4	2.6	2.0	1.7
			250	5.3	3.7	2.7	2.1	1.6	1.3
	Deflection	in Inches.	... ..	.23	.33	.44	.58	.74	.90
14	7	52	100	11.5	8.0	5.9	4.5	3.5	2.9
			150	7.7	5.3	3.9	3.0	2.4	1.9
			200	5.7	4.0	2.9	2.2	1.8	1.4
			250	4.6	3.2	2.3	1.6	1.4	1.1
	Deflection	in Inches.	......	.23	.33	.44	.58	.74	.90

PENCOYD BEAMS.

Figures in small type denote that the beams so placed will deflect more than $\frac{1}{30}$ of an inch for each foot of span.

Length of Span in Feet.						Load per Sq. Ft. of Floor, Lbs.	Weight per Yard, Lbs.	Size of Beam in Inches.	Chart Number.
22	24	26	28	30	32				
Distance between Centres of Beams in Feet.									
6·7	5·6	4·8	4·1	3·6		100			
4·5	3·7	3·2	2·8	2·4		150			
3·3	2·8	2·4	2·1	1·8	'	200	112	10	7
2·7	2·3	1·9	1·6	1·4		250			
·76	·92	1·06	1·26	1·44	......	......	Deflection	in Inches.	
5·7	4·5	4·1	3·5	3·1		100			
3·8	3·2	2·7	2·4	2·0		150			
2·9	2·4	2·0	1·8	1·5		200	90	10	8
2·3	1·9	1·6	1·4	1·2		250			
·78	·92	1·08	1·26	1·44	......	......	Deflection	in Inches.	
5·1	4·3	3·6	3·1			100			
3·4	2·8	2·4	2·1			150			
2·5	2·1	1·8	1·6			200	90	9	9
2·0	1·7	1·5	1·3			250			
·86	1·02	1·21	1·40	......	......	......	Deflection	in Inches.	
4·0	3·4	2·9	2·5			100			
2·7	2·3	2·0	1·7			150			
2·0	1·7	1·4	1·2			200	70	9	10
1·6	1·4	1·2	1·0			250			
·86	1·02	1·21	1·40	......	......	......	Deflection	in Inches.	
4·0	3·4	2·9				100			
2·7	2·3	1·9				150			
2·0	1·7	1·4				200	81	8	11
1·6	1·4	1·2				250			
·97	1·16	1·36	......	......	......	......	Deflection	in Inches.	
3·3	2·8	2·4				100			
2·2	1·9	1·6				150			
1·7	1·4	1·2				200	65	8	12
1·3	1·1	1·0				250			
·97	1·16	1·36	......	......	......	......	Deflection	in Inches.	
2·7	2·3					100			
1·6	1·5					150			
1·4	1·2					200	65	7	13
1·1	·9					250			
1·09	1·32	......	......	......	......	......	Deflection	in Inches.	
2·4	2·0					100			
1·6	1·3					150			
1·2	1·0					200	52	7	14
·9	·8					250			
1·09	1·32	......	......	......	......	......	Deflection	in Inches.	

PENCOYD  BEAMS.

Greatest distance between centres of floor beams so that the bending stress on beam will not exceed its maximum safe load.

Figures in small type denote that the beams so placed will deflect more than $\frac{1}{30}$ of an inch for each foot of span.

Chart Number.	Size of Beam in Inches.	Weight per Yard, lbs.	Load per sq. ft. of Floor, lbs.	Length of span in feet.					
				10	12	14	16	18	20
				Distance between centres of beams in feet.					
15	6	50	100	8.4	5.8	4·2	3·3	2·6	2·1
			150	5.6	3.9	2·8	2·2	1·7	1·4
			200	4.2	2.9	2·1	1·6	1·3	1·0
			250	3.3	2.3	1·7	1·3	1·0	·8
	Deflection	in Inches.	......	.27	.38	·52	·69	·87	1·07
16	6	40	100	7.5	5.2	3·8	2·9	2·3	1·9
			150	5.0	3.5	2·5	1·9	1·5	1·2
			200	3.7	2.6	1·9	1·6	1·2	·9
			250	3.0	2.1	1·5	1·2	·9	·7
	Deflection	in Inches.	......	.27	.38	·52	·69	·87	1·07
17	5	34	100	5.0	3·5	2·6	1·9	1·5	1·2
			150	3.3	2·3	1·7	1·3	1·0	·9
			200	2.5	1·7	1·3	1·0	·8	·6
			250	2.0	1·4	1·0	·8	·6	·5
	Deflection	in Inches.	......	.32	·46	·63	·82	1·04	1·29
18	5	30	100	4.7	3·2	2·4	1·8	1·4	1·2
			150	3.1	2·1	1·6	1·2	·9	·8
			200	2.3	1·5	1·2	·9	·7	·6
			250	1.9	1·3	1·0	·7	·6	·5
	Deflection	in Inches.	......	.32	·46	·63	·82	1·04	1·29
19	4	28	100	3·6	2·5	1·8	1·4	1·1	
			150	2·4	1·7	1·2	·9	·7	
			200	1·8	1·2	·9	·7	·6	
			250	1·4	1·0	·7	·6	·4	
	Deflection	in Inches.	......	·40	·58	·79	1·03	1·31	
20	4	18.5	100	2·4	1·7	1·2	·9	·7	
			150	1·6	1·1	·8	·6	·5	
			200	1·2	·8	·6	·5	·4	
			250	1·0	·7	·6	·4	·3	
	Deflection	in Inches.	......	·40	·55	·79	1·03	1·31	
21	3	23	100	2·0	1·4	1·0	·8		
			150	1·3	·9	·7	·5		
			200	1·0	·7	·5	·4		
			250	·8	·6	·4	·3		
	Deflection	in Inches.	......	·53	·77	1·05	1·37		
22	3	17	100	1·6	1·1	0·8	·6		
			150	1·1	·7	·5	·4		
			200	·8	·6	·4	·3		
			250	·6	·4	·3	·2		
	Deflection	in Inches.	......	·53	·77	1·06	1·37		

TIE RODS FOR BEAMS SUPPORTING BRICK ARCHES.

The horizontal thrust of Brick arches is found as follows:

$$\frac{1.5 \; WL^2}{R} = \text{pressure in lbs. per lineal foot of arch.}$$

$W =$ Load in lbs. per square foot.

$L =$ Span of arch in feet

$R =$ Rise in inches.

Place the tie rods as low through the webs of the beams as possible, and spaced so that the pressure of arches as obtained above will not produce a greater stress than 15,000 lbs. per square inch of the least section of the bolt.

Example.—The beams supporting an arched brick floor are five feet apart, and the rise of the arches is six inches. The total weight of floor and load equals 150 lbs. per square foot.

Then $\dfrac{1.5 \times 150 \times 25}{6} = 937.5$ lbs. pressure per lineal foot of

arch. If one-inch screw bolts are used which have an effective section of $\frac{6}{10}$ square inches. Then $.6 \times 15,000 = 9,000$ lbs. which is the greatest load the bolt should be allowed to sustain, and

$\dfrac{9,000}{937.5} = 9.6$ feet $=$ greatest distance apart of the bolts, or in

same manner we would find 5.3 feet, if $\frac{7}{8}$ inch tie rods are used.

Ordinarily it will be found necessary to limit the spacing of the tie rods to avoid excessive bending stress on the outer beams of the floor, or to prevent this bending stress being transferred to the walls of the building.

The ability of the outer beams to resist the horizontal bending action caused by the pressure of the arches is determined as follows:

LATERAL STRENGTH OF FLOOR BEAMS.

The resistance to bending of any **I** Beam or Channel bar, for a force acting at right angles to the web, or in the direction of the flanges,

$$W = \frac{10\,I}{LF} \text{ for } \textbf{I} \text{ Beams.}$$

$$W = \frac{8\,I}{LF} \text{ for Channels.}$$

$W =$ Safe distributed load in net tons.
$L =$ Length in feet between supports.
$F =$ Width of flange in inches.
$I =$ Moment of inertia, axis coincident with web, see col. viii., pages 92–101.

The above gives results which have been proved by experiment not to exceed one-third the ultimate strength of the beams. The formulæ given properly apply to beams secured at each end only. If the beam is of considerable length requiring supports at several points, it can be considered as *continuous* (see page 75), and the formulæ become,

$$W = \frac{15I}{LF}, \text{ for I Beams.}$$

$$W = \frac{12I}{LF}, \text{ for Channels.}$$

Example.—A 9-inch 70 lb. I Beam forming the outer support for an arched brick floor has the tie rods at intervals of 6 feet. What evenly distributed horizontal pressure will it safely resist ? $I = 5.6$ (see col. viii., page 92). $F = 4\frac{1}{2}$ inches (see col. C, page 2). Then $W = \dfrac{15 \times 5.6}{6 \times 4\frac{1}{2}} = 3.4$ tons or 1,130 lbs. per lineal foot of arch.

Knowing the amount of the load W and requiring the distance L. Above equation becomes $L^2 = \dfrac{15I}{W^1 F}$ in which $W^1 =$ pressure or load on beam per lineal foot.

Example.—An 8″ 43 lb. channel bar forms the end support for a system of brick arches having a span of 4 feet and 4 inches rise. How closely ought tie rods to be placed so that the channels will not be overstrained ? The horizontal thrust per lineal foot of arch $= \dfrac{1.5 \times 150 \times 16}{4} = 900$ lbs. or .45 tons. $I = 2.17$. $F = 2\frac{9}{32}$.

$$L^2 = \frac{12 \times 2.17}{.45 \times 2\frac{9}{32}} \text{ or } L = 5 \text{ feet.}$$

It will generally be found that an *angle* bar makes a better and more economical support for the arches on the side walls than either an I beam or channel.

The resistance to bending of an angle is readily found by the rule given on page 69.

$$W = \frac{.93AD}{L} = \text{safe distributed load for a non-continuous}$$

beam.

$$W = \frac{1.4AD}{L} = \text{safe distributed load for a continuous beam.}$$

And as before $L^2 = \frac{1.4AD}{W}$. A being the sectional area in square inches, and D the width or size of the angle in inches.

Applying this rule to the last example, and considering the 8" channel replaced by a 4" × 4" × ½" angle whose area = 3.75 square inches.

$$L^2 = \frac{1.4 \times 3.75 \times 4}{.45} = 46.6 \text{ or } L = 6.8 \text{ feet between centers}$$

of bolts. Stress on bolts 900 × 6.8 = 6,120 lbs. To resist this ⅞" would be the proper diameter of the screw.

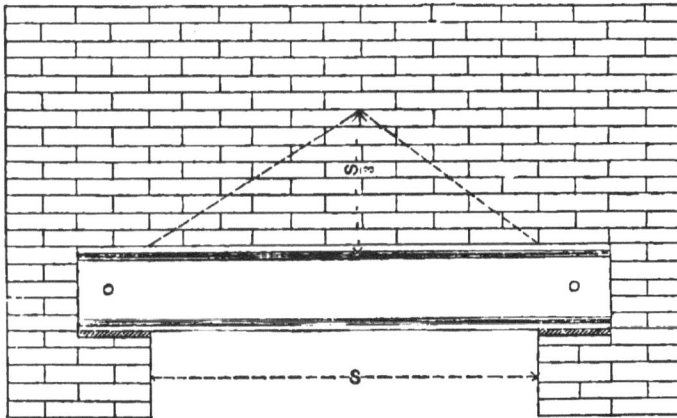

BEAMS SUPPORTING BRICK WALLS.

If the wall has no openings and the bricks are laid with the usual bond, the prism of wall that the beam sustains will be of a triangular shape, the height being one-fourth of the span. Owing to frequent irregularities in the bonding, it is best to consider the height as one-third of the span.

The weight of brick work for each inch of thickness, is about 10 lbs. per square foot. Therefore the weight of the triangular mass of brick that the beam supports is found as follows :

$$\frac{span \times \frac{span}{3} \; in \; feet}{2} \times 10 \; times \; the \; thickness \; of \; the \; wall \; in \; inches$$

= weight in lbs.; or reducing above to its more concise form,

$$W = \frac{10ts^2}{6}.$$

$W =$ Weight in lbs. supported by the beam.

$t =$ Thickness of wall in inches.

$s =$ Span of beam in feet.

The greatest bending stress at the center of the beam, resulting from a brick wall of above shape, is the same as that caused by a load one-sixth less concentrated at the center of the beam.

Example.—What beam will be required to span an opening of 16 feet, and carry a solid brick wall 8 inches thick, the beam not to be strained more than one-third of its ultimate strength ?

Weight of wall by the rule. $W = \dfrac{10 \times 8 \times 256}{6} = 3{,}413 \; lbs.$

Considering the load as in middle of beam, it would be five-sixths of above = 2,845 lbs., or 5,690 lbs. if evenly distributed.

By our table page 43, a 7″ I beam 52 lbs. per yard, comes nearest to what is required, its greatest safe distributed load being 3.5 tons. The deflection under this load will be about .45 of an inch, found as described on page 89.

If a wall has openings such as windows, etc.. the imposed weight on the beam may be greater than if the wall is solid.

For such a case consider the outline of the brick, which the beam sustains, to pass from the points of support diagonally to the outside corners of the nearest openings, then vertically up the outer line of the jambs, and so on if other openings occur above. If there should be no other openings, consider the line of imposed brick work to extend diagonally up from each upper corner of the jambs, the intersection forming a triangle whose height is one-third of its base, as described at beginning.

APPROXIMATE FORMULÆ FOR ROLLED IRON BEAMS.

The following rules for the strength and stiffness of rolled iron beams of various sections are intended for convenient application in cases where strict accuracy is not required.

The rules have been derived from the authoritative formulæ. Those for rectangular and circular sections are correct, while those for the flanged sections are limited in their application to the standard shapes as given in our tables. They will be found to give results which have been proved by experiment to be sufficiently accurate for practical purposes. When the section of any beam is increased above the standard minimum dimensions, the flanges remaining unaltered, and the web alone being thickened, the tendency will be for the ultimate load as found by the rules to be in excess of the actual, but within the limits that it is possible to vary any section in the rolling, the rules will apply without any serious inaccuracy.

IN THE TABLES OF FORMULÆ

Column   I. indicates the cross section of the beam.

Column II. gives the ultimate load applied at the center of a beam supported at each end.

Column III. gives the ultimate load uniformly distributed over a beam supported at each end.

Column IV. indicates the deflection under any load, w (not exceeding one-half the ultimate load) at the middle of the beam.

Column V. gives the deflection for a load uniformly distributed.

SAFE LOADS.

The ultimate load given in the tables is defined on page 32. One-third of this should be accepted as the greatest safe stationary load, and from one-fourth to one-sixth of the same when a moving or fluctuating load is imposed, according to the way it is applied, or the degree of stiffness required. See table, page 34.

10 A = WEIGHT PER YARD IN LBS.

The area, A, of any cross section of wrought iron may be obtained by dividing its weight per yard by 10 ; and *vice versa*, its weight per yard may be found by multiplying its area in square inches by 10 ; *e.g.* the area of a beam weighing 50 lbs. per yard is five square inches.

TABLE OF FORMULÆ FOR WROUGHT IRON BEAMS.

For greatest safe load take one-third of the ultimate as obtained below.

SHAPE OF SECTION.	W — ULTIMATE LOAD IN NET TONS.		Δ — DEFLECTION IN INCHES.	
COLUMN I.	IN MIDDLE. II.	DISTRIBUTED. III.	LOAD IN MIDDLE. IV.	DISTRIBUTED LOAD. V.
SOLID RECTANGLE.	1. $$W = \frac{1.3\,AD}{L}$$	2. $$W = \frac{2.6\,AD}{L}$$	3. $$\Delta = \frac{wL^3}{30\,AD^2}$$	4. $$\Delta = \frac{wL^3}{48\,AD^2}$$
HOLLOW RECTANGLE.	5. $$W = \frac{1.3\,(AD-ad)}{L}$$	6. $$W = \frac{2.6\,(AD-ad)}{L}$$	7. $$\Delta = \frac{wL^3}{(30\,AD^2-ad^2)}$$	8. $$\Delta = \frac{wL^3}{48\,(AD^2-ad^2)}$$
SOLID CYLINDER.	9. $$W = \frac{0.95\,AD}{L}$$	10. $$W = \frac{1.9\,AD}{L}$$	11. $$\Delta = \frac{wL^3}{23\,AD^2}$$	12. $$\Delta = \frac{wL^3}{37\,AD^2}$$
HOLLOW CYLINDER.	13. $$W = \frac{0.95\,(AD-ad)}{L}$$	14. $$W = \frac{1.9\,(AD-ad)}{L}$$	15. $$\Delta = \frac{wL^3}{23\,(AD^2-ad^2)}$$	16. $$\Delta = \frac{wL^3}{37\,(AD^2-ad^2)}$$

Even-legged Angle or Tee. 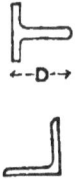	17. $$W = \frac{1.4\,AD}{L}$$	18. $$W = \frac{2.8\,AD}{L}$$	19. $$\Delta = \frac{wL^3}{34\,AD^2}$$	20. $$\Delta = \frac{wL^3}{54\,AD^2}$$
Channel Bar. 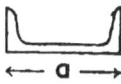	21. $$W = \frac{1.9\,AD}{L}$$	22. $$W = \frac{3.8\,AD}{L}$$	23. $$\Delta = \frac{wL^3}{50\,AD^2}$$	24. $$\Delta = \frac{wL^3}{80\,AD^2}$$
Deck Beam. 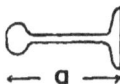	25. $$W = \frac{2\,AD}{L}$$	26. $$W = \frac{4\,AD}{L}$$	27. $$\Delta = \frac{wL^3}{52\,AD^2}$$	28. $$\Delta = \frac{wL^3}{83\,AD^2}$$
I Beam. 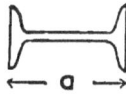	29. $$W = \frac{2.1\,AD}{L}$$	30. $$W = \frac{4\,2\,AD}{L}$$	31. $$\Delta = \frac{wL^3}{56\,AD^2}$$	32. $$\Delta = \frac{wL^3}{90\,AD^2}$$

L = Length in feet between supports.
A = Sectional area of beam in square inches.
D = Depth of beam in inches.

a = Interior area in square inches.
d = Interior depth in inches.
w = Working load in net tons.

EXAMPLES CALCULATED FROM PRECEDING TABLES.

SOLID RECTANGULAR SECTIONS.

Example 1.—To find the breaking load for any solid rectangular beam loaded in the middle.

C = Solid rectangular bar, 2 inches wide, 4 inches deep and 10 feet between supports. Then, from Formula No. 1, we have $\dfrac{1.3 \times 8 \times 4}{10} = 4.16$ tons breaking load in middle of beam.

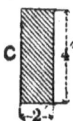

Example 2.—To find the uniformly-distributed breaking load for same beam.

Formula No. 2. $\dfrac{2.6 \times 8 \times 4}{10} = 8.32$ tons breaking load uniformly distributed.

Example 3.—To find the deflections for above beam under the greatest safe loads; viz., one-third breaking loads.

Formula No. 3. $\dfrac{1.39 \times 1000}{30 \times 8 \times 16} = 0.36$ inches, for a load of 1.39 tons in middle.

Formula No. 4. $\dfrac{2.77 \times 1000}{48 \times 8 \times 16} = 0.45$ inches, for a load of 2.77 tons distributed.

HOLLOW RECTANGULAR SECTIONS.

Example 4.—To find the breaking loads for any hollow rectangular beam supported at both ends.

Let C be a hollow rectangular section, 4 inches wide, 8 inches deep, external dimensions ; 3 inches wide, 6 inches deep, internal dimensions; 15 feet between supports.

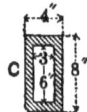

Formula No. 5. $\dfrac{1.3 \left[(32 \times 8) - (18 \times 6) \right]}{15} = 12.83$ tons, breaking load in middle; and multiplying this result by 2, we have 25.66 tons for the breaking load uniformly distributed.

Example 5. To find the deflection of this beam with three tons in middle; also with six tons distributed.

Formula No. 7. $\dfrac{3 \times 3375}{30\left[(32 \times 64)-(18 \times 36)\right]} = 0.24$ inches deflection with three tons in middle.

Formula No. 8. $\dfrac{6 \times 3375}{48\left[(32 \times 64)-(18 \times 36)\right]} = 0.3$ inches deflection with six tons distributed.

SOLID AND HOLLOW CYLINDERS.

The preceding examples for rectangles will apply to the circular sections by merely substituting the proper co-efficients as given in Formulæ 9 to 16 inclusive.

EVEN-LEGGED ANGLES AND TEES.

Example 6.—To find the breaking loads for an even-legged angle or tee, used as a beam supported at both ends.

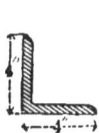

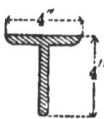

Weight, 37 lbs. per yard or 3.7 square inches section; 12 ft. between supports.

Formula No. 18. $\dfrac{2.8 \times 3.7 \times 4}{12} = 3.45$

tons breaking load uniformly distributed, or 1.73 tons breaking load in the middle.

Example 7.—To find the deflection of the above beam under a load suspended from the middle of the beam.

Load = 1500 lbs. = .75 tons.

Formula No. 19. $\dfrac{.75 \times 1728}{34 \times 3.7 \times 16} = .64$ inches deflection.

Theoretically an angle has the same transverse strength as a tee of the same dimensions. But owing to the difficulty of disposing the load as symmetrically on the angle as on the tee, the latter shape generally yields better results by experiment.

CHANNEL BARS.

Example 8.—To find the breaking loads for a channel bar used as a beam supported at both ends.

Channel bar 9 inches deep, 70 pounds per yard; 7 square inches section, 14 feet between supports.

Formula No. 22. $\dfrac{3.8 \times 7 \times 9}{14} = 17.1$ tons distributed

breaking load, or half this weight will be the breaking load in the middle.

Example 9.—To find the deflection of above beam under greatest safe distributed load.

$\dfrac{17.1}{3} = 5.7$ tons greatest safe distributed load.

Formula No. 24. $\dfrac{5.7 \times 2744}{80 \times 7 \times 81} = 3.5$ inches deflection.

I BEAMS.

Example 10.—To find the breaking loads for an I beam, loaded in the middle and supported at both ends.

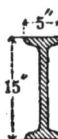

A 15″ I beam, 200 lbs. per yard, 20 square inches area, 20 feet between supports. Formula No. 29. $\dfrac{2.1 \times 20 \times 15}{20}$

$= 31.5$ tons middle breaking load; one-third of which (10.5 tons) will be greatest safe load in middle, or twice this (21 tons) equals greatest safe load distributed.

Example 11.—To find the deflections for the same I beam under the above greatest safe loads.

Formula No. 31. $\dfrac{10.5 \times 8000}{56 \times 20 \times 225} = .33$ inches under a load of 10.5 tons in the middle.

Formula No. 32. $\dfrac{21 \times 8000}{90 \times 20 \times 225} = .41$ inches under a load of 21 tons uniformly distributed.

Although the preceding rules for I beams and channels give results which are substantially correct for all the standard sections as ordinarily rolled, yet they are not strictly accurate, and not applicable to the heavier-built beams, whose flanges are much larger, relatively to the web, than is the case in the average rolled beams. For such cases, the following formula is correct. $\dfrac{6.6\,A'\,D' + 1.2\,a'd'}{L} =$ breaking load in middle of beam.

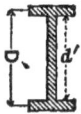

$A' =$ Area of one flange.
$D' =$ Depth between centres of flanges.
$a' =$ Area of web.
$d' =$ Depth of web.

For example, a beam 20 inches deep, flanges $8'' \times 1''$, web $\frac{1}{4}''$ thick, 20 feet between supports,

$$\frac{6.6 \times 8 \times 19'' + 1.2 \times 4.5 \times 18}{20} = 55 \text{ tons}$$

breaking load in middle of beam; whereas the Rule in Table for Rolled Beams would give a similarly placed load of

$$\frac{2.1 \times 20.5 \times 20}{20} = 43 \text{ tons.}$$

When the load is concentrated away from the centre of beam, the ultimate load will be to the load at centre as the square of half the span is to the product of the segments formed by position of load.

Example.—A beam 20 feet between supports has its load placed 5 and 15 feet respectively from each end : the breaking load at that point is to the calculated breaking centre load as 100 is to 75.

BEAMS HAVING NO LATERAL SUPPORT BETWEEN BEARINGS.

If beams are used without any support sideways, the tendency to fail, by lateral bending of the top flange, will increase with the length of the beam; and, in such cases, it is better to limit the application of the preceding rules to beams whose lengths do not exceed 20 times the width of the flange, gradually increasing the factor of safety for longer beams ; so that, when the beam reaches a length equal to 70 times the width of the flange, the greatest safe load would be about one-sixth of the calculated breaking load, or the proper factor of safety for the latter beam would be double that for the former. (See page 36.)

CANTILEVER BEAMS.

The application of the preceding rules to overhanging beams fixed at one end and free at the other, is best indicated by supposing a beam with both ends supported to be inverted, and the reaction of the supports considered as the positive load.

It is then evident that a beam, $A\,C$ (see above illustration), both ends supported, will be strained with a middle load, W, in an equal manner to a cantilever, $A\,B$ or $B\,C$, of half the length of $A\,C$ and having a similar section, and bearing one-half the load $\left(\text{or}\,\dfrac{W}{2}\right)$ at its end.

EXAMPLES FOR CANTILEVER BEAMS.

A rectangular bar, $6'' \times 2''$, built into a wall and projecting eight feet. For load concentrated at its end, take one-fourth the co-efficient in Table for Beams with both ends supported and load in middle. $\dfrac{1.3 \times 12 \times 6}{4 \times 8} = 2.9$ tons ultimate load. Deflection under one-third of above, or say nine-tenths of a ton; substituting one-sixteenth of the co-efficient for deflection when load is in middle. $\dfrac{9 \times 512}{1\frac{3}{4} \times 12 \times 36} = 0.56$ inches deflection at end.

A 12-inch I beam, 15 square inches section, extends 10 feet beyond a rigid support. For a load evenly distributed, take one-fourth the co-efficient for a beam supported at both ends, bearing a distributed load. $\dfrac{1.05 \times 15 \times 12}{10} = 18.9$ tons breaking load distributed.

For deflection under five tons distributed, substitute one-sixth of the co-efficient for deflection in Rule for Beams supported at both ends with load in middle. $\dfrac{5 \times 1000}{9.33 \times 15 \times 144} = 0.25$ inches deflection at end of beam.

CONTINUOUS BEAMS.

When a beam is continuous over several supports, or when both ends are as rigidly secured as is necessary at the fixed ends of a cantilever, the beam is practically in the same condition as a non-continuous beam of shorter span.

When the load is applied at the middle of the span, the ultimate breaking load of a continuous beam is equal to twice that for a non-continuous beam similarly loaded and of the same length and section.

When the load is evenly distributed, the ultimate load for a continuous beam is 1.5 times greater than the ultimate load for a non-continuous beam under the same conditions and of the same length and section.

The deflection of a continuous beam is one-fourth that of a non-continuous beam when similarly loaded.

To find the strength and stiffness of continuous beams, take the rules given for non-continuous beams and alter the co-efficients in the proportions stated.

EXAMPLES FOR CONTINUOUS BEAMS.

A 4-inch I beam of three square inches section is continuous over supports twenty feet apart. To find the greatest safe load uniformly distributed, and corresponding deflection, take 1.5 times the co-efficient for a similar non-continuous beam. $\dfrac{6.3 \times 3 \times 4}{20} = 3.78$ tons breaking load, or 1.26 tons safe distributed load. For deflection, take four times the co-efficient for the same class of non-continuous beam. $\dfrac{1.26 \times 8000}{360 \times 3 \times 16} = 0.58$ of an inch deflection.

For a continuous beam bearing load in middle, take twice the co-efficient given for the strength of a similarly loaded non-continuous beam, and, for deflection of the former, take four times the co-efficient given for the latter beam.

It will be observed that these rules apply only to the intermediate spans of continuous beams, as, owing to the failure of continuity at one end of each outer span, the conditions are altered. If, however, the outer ends of a continuous beam overhang the end-supports from one-fifth to one-fourth of a span, and bear the same proportion of load as the parts between supports, then the outer spans may be of same length as the intermediate spans, subject to the same load, and the strength and stiffness are determined by the same rules ; otherwise, the outer spans ought to be only four-fifths of the length of the intermediate spans when the load is distributed, or three-fourths of the same when the load is concentrated in the middle ; or, if the lengths of spans are all alike, the loads on outer spans ought to be reduced in the same proportion.

The following table exhibits the relative proportions of strength and stiffness existing between the various classes of beams when they have the same lengths and uniform cross sections ; the deflections being comparative figures for the same loads.

KIND OF BEAM.	Breaking load as	Deflection as
Fixed at one end—loaded at the other.......	$\frac{1}{4}$	16
Fixed at one end—load evenly distributed...	$\frac{1}{2}$	6
Supported at both ends—load in middle.....	1	1
Supported at both ends—load evenly distributed......................................	2	$\frac{1}{8}$
Continuous beam—load in middle...........	2	$\frac{1}{4}$
Continuous beam—load evenly distributed...	3	$\frac{5}{32}$

The breaking load and deflection of a beam supported at both ends and loaded in the middle have been taken as the units in the preceding table, and—the proportional strength and stiffness of similar beams under different conditions given—to find the proper co-efficient for estimating the strength and stiffness of the beam required, it is necessary to alter, in the given proportions, the co-efficient for the same beam when supported at both ends and loaded in the middle.

CHANGES OF CO-EFFICIENTS FOR SPECIAL FORMS OF BEAMS.

For beams of the character denoted in list below, change the co-efficients in table of formulæ, pages 68–69, in the ratio given. For concentrated loads and distributed loads respectively, change the co-efficients given for the same kinds of loads in the table.

KIND OF BEAM.	CO-EFFICIENT FOR ULTIMATE LOAD.	CO-EFFICIENT FOR DEFLECTION.
Fixed at one end, loaded at the other.	One-fourth ($\frac{1}{4}$) of the co-efficient of table.	One - sixteenth ($\frac{1}{16}$) of the co-efficient of table.
Fixed at one end, load evely distributed.	One-fourth ($\frac{1}{4}$) of the co-efficient of table.	Five - forty-eighths ($\frac{5}{48}$) of the co-efficient of tables.
Both ends rigidly fixed, or a continuous beam, with load in middle.	Twice the co-efficient of table.	Four times the co-efficient of table.
Both ends rigidly fixed, or a continuous beam, with load evenly distributed.	One and one-half ($1\frac{1}{2}$) times the co-efficient of table.	Four times the co-efficient of table.

BENDING MOMENTS AND DEFLECTIONS FOR BEAMS OF UNIFORM SECTION.

W = Total load. E = Modulus of elasticity.
L = Length of beam. I = Moment of inertia.

FORM OF BEAM AND POSITION OF LOAD.	Maximum bending moment.	Maximum shearing stress.	Deflection.
Beam fixed at one end loaded at the other : **FIG. 1** 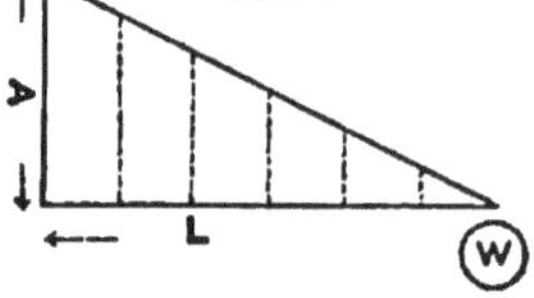 Draw triangle having $A = WL$. Vertical lines give bending moments at corresponding points on the beam.	at point of support $= WL$.	at point of support $= W$.	at end of beam $= \dfrac{WL^3}{3EI}$.
Beam fixed at one end, load uniformly distributed : **FIG. 2** 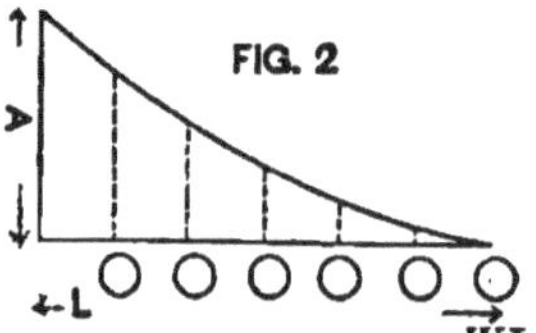 Draw parabola having $A = \dfrac{WL}{2}$ Ordinates give bending moments at corresponding points on the beam.	at point of support $= \dfrac{WL}{2}$.	at point of support $= W$.	at end of beam $= \dfrac{WL^3}{8EI}$.
Beam supported at both ends, loaded in the middle : **FIG. 3**  Draw triangle having $A = \dfrac{WL}{4}$. Vertical lines give bending moments at corresponding points on the beam.	at middle of beam $= \dfrac{WL}{4}$.	at point of support $= \dfrac{W}{2}$.	at middle of beam $= \dfrac{WL^3}{48EI}$.

BENDING MOMENTS AND DEFLECTIONS FOR BEAMS OF UNIFORM SECTION.

$W =$ Total load.	$E =$ Modulus of elasticity.
$L =$ Length of beam.	$I =$ Moment of inertia.

FORM OF BEAM AND POSITION OF LOAD.	Maximum bending moment.	Maximum shearing stress.	Deflection.
Beam supported at both ends, load uniformly distributed : FIG.4 Draw parabola having $A = \dfrac{WL}{8}$. Ordinates give bending moments at corresponding points on the beam.	at middle of beam $= \dfrac{WL}{8}$.	at point of support $= \dfrac{W}{2}$.	at middle of beam $= \dfrac{WL^3}{76.8EI}$.
Beam supported at both ends, load concentrated at any point : FIG. 5 Draw triangle having $A = \dfrac{Wab}{L}$. Vertical lines give bending moments at corresponding points on the beam.	at position of load $= \dfrac{Wab}{L}$.	at point of support next to a $= \dfrac{Wb}{L}$. at point of support next to b $= \dfrac{Wa}{L}$.	at position of load $= \dfrac{a^2b^2 W}{3EIL}$.

BENDING MOMENTS AND DEFLECTIONS FOR BEAMS OF UNIFORM SECTION.

W = Total load. E = Modulus of elasticity.
L = Length of beam. I = Moment of inertia.

Beam supported at both ends, with concentrated load at various points :

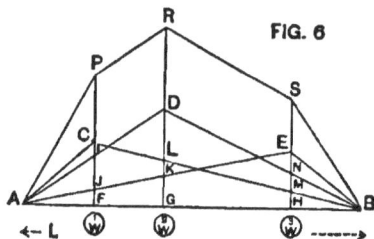

FIG. 6

Draw (by 5) the triangles having vertices at C, D and E, the verticals representing bending moments for loads w^1, w^2 and w^3, respectively. Extend FC to P, GD to R, and HE to S, making each long vertical equal to the sum of the bending moments corresponding to its position. That is, $FP = FC + FI + FJ$. $GR = GD + GL + GK$. And $HS = HE + HN + HM$. Verticals drawn from any point on the polygon, $APRSB$ to AB, will represent the bending moments at the corresponding points on the beam.

Beam rigidly secured at each end, and loaded in the middle. Or the intermediate spans of a continuous beam, equally loaded in the middle of each span :

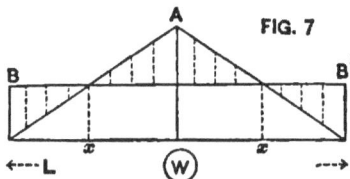

FIG. 7

Points of contraflexure at x, x, where Moment = 0. Distance of x from either support = $\dfrac{L}{4}$. Equal moments at middle and ends = $\dfrac{WL}{8}$. Deflection $= \dfrac{WL^3}{192EI}$

Draw a triangle having $A = \dfrac{WL}{4}$, and at ends draw verticals BB', each $= \dfrac{WL}{8}$, join BB'. The vertical distances between BB' and the sides of the triangle, represent the moments for corresponding points cn the beam.

BENDING MOMENTS AND DEFLECTIONS FOR BEAMS OF UNIFORM SECTIONS.

W = Total load.	E = Modulus of elasticity.
L = Length of beam.	I = Moment of inertia.

Beam rigidly secured at each end with load uniformly distributed.

Or the intermediate spans of a continuous beam bearing a uniformly distributed load on each span :

FIG. 8

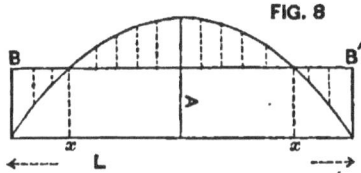

Points of contraflexure x, x, where moment = 0. Distance of x from either support = .21L.

Draw parabola having $A = \dfrac{WL}{8}$. Draw verticals B, B', each equal to $\dfrac{WL}{12}$, join BB'. The vertical distances between BB' and the curve of the parabola represent the moments for corresponding points on the beam.

Maximum moment at points of support = $\dfrac{WL}{12}$.

Moment at middle of beam = $\dfrac{WL}{24}$.

Maximum deflection at middle of beam = $\dfrac{WL^3}{307.2EI}$

6

BEAMS FOR SUPPORTING IRREGULAR LOADS.

When a beam has its load unequally distributed over it, the proper size of the beam can be determined by finding the maximum bending moment and proportioning the beam accordingly. Equilibrium is obtained when the bending moment is equal to the moment of resistance. That is, when the external force multiplied by the leverage with which it acts is equal to the strength of the material in the cross section of the beam multiplied by the leverage with which it acts. The ultimate moment of resistance for a wrought-iron beam of symmetrical form is

$$\frac{42000 \, I}{\frac{1}{4} \, \text{depth}} \quad \text{or} \quad \frac{84000 \, I}{d}.$$

$d =$ depth of beam in the direction in which the force acts.
$I =$ the moment of inertia about the axis at right angles to the direction of the force.

The greatest safe moment of resistance as adopted in our tables is one-third ($\frac{1}{3}$) of above,

$$M = \frac{28000 \, I}{d} \quad \text{or} \quad \frac{M}{28000} = \frac{I}{d}.$$

The co-efficient to be changed according to the factor of safety desired. The rule would thus be $\dfrac{\text{Moment}}{\text{Co-efficient}} = \dfrac{I}{d}$.

RULE FOR BEAMS BEARING IRREGULAR LOADS.

Find by the methods described in preceding article the maximum bending moment in inch-lbs. for the loads. Divide the moment by the proper co-efficient as described above. Find in the tables, pages 92-96, a beam whose inertia divided by its depth is not less than this quotient; which will be the beam required.

In some instances the maximum bending moment can be most readily found by the use of diagrams, as described in the succeeding article.

When this is done use any convenient scale, making all loads and all distances respectively of the same denominations.   The maximum bending moment can then be measured to scale.

Example.—An **I** beam 8 feet long is to be fixed at one end and loaded at the other with 5,000 lbs. and carrying also an evenly distributed load of 8,000 lbs.   What size of beam should be used so as not to be strained over one-third of its ultimate capacity ?

Moment for end load $\qquad$ = 5,000 × 96 = 480,000 inch-lbs.

"   "   distributed load $= \dfrac{8,000 \times 96}{2} = 384,000$ "

$$\text{Total} = 864,000 \quad \text{"}$$

For one-third ($\frac{1}{3}$) of ultimate the co-efficient will be

$$\frac{84,000}{3} = 28,000.$$

$$\frac{864.000}{28,000} = 30.84 = \frac{I}{d}.$$

By Column VII., page 92, for a 12″ 168 lb. **I** beam, $I =$ 371.98, which divided by 12 = 30.99; or a 15″ 145 lb. **I** beam, $\dfrac{I}{d} = 34.7$.   The latter beam would be stronger and lighter.

In the following example the maximum bending moment can be very readily obtained by a diagram as described in Fig. 6 of the preceding article.

Example.—A beam 20 feet long between supports, will carry three loads, which we will call *A*, *B*, and *C*.

$A = 4,000$ lbs. and is 4 feet from one end of the beam.
$C = 6,000$ lbs. and is 3 feet from the other end of the beam.
$B = 5,000$ lbs. and is 5 feet from *C* and 8 feet from *A*.

What beam is best to use for above, not strained over one-fourth of the ultimate ?   Describe the diagram as per Fig. 6, when the following bending moments in ft.-lbs. will be obtained.

At point A	At point B	At point C
For load A.. 12,800	For load B.. 24,000	For load C.. 15,300
" B.. 8,000	" A.. 10,800	" B.. 8,900
" C.. 3,600	" C.. 6,400	" A.. 2,400
Total.....24,400	Total.... 41,200	Total.... 26,600

The maximum moment at $B = 41,200$ ft.-lbs. or 494,400 inch-lbs. For one-fourth of ultimate strength co-efficient $= 21,000$.

$$\frac{494,400}{21,000} = 23.5 = \frac{I}{d}.$$

By table on page 92, for a 12″ 120 lb. I beam $\frac{I}{d} = 22.74$, being slightly deficient. A 12″ 125 lb. I beam will be ample.

If more lateral stiffness is required than a single beam affords, use a pair of channels separated and braced horizontally. Two 12″ 75 lb. channels $\frac{I}{d} = 23.6$, would suit above purposes.

NOTE.—The tables of elements, except where otherwise specified, are calculated for dimensions in inches and weights in lbs., consequently in examples of above character it is necessary to obtain bending moments in inch-lbs.

BEAMS SUBJECT TO BOTH BENDING AND COMPRESSION.

When a beam is subjected to bending action and simultaneously has to act as a strut by resisting compression, the stress of the fibres of the beam in tension will be relieved and those in compression correspondingly augmented.

No general rules can be given for such conditions, as every particular case requires its own proper determination. The following methods, though not strictly correct, will give safe results for some simple forms of trussed girders, etc.

(1.) When the beam is subject to compression but is so confined laterally that it cannot fail by bending like a strut.

Rule.—Find the section of beam required to resist bending, then allowing from 10,000 to 15,000 lbs. per square inch of section for the compression, according to the factor of safety used, add the area so found to the first area, which will give the section of required beam.

Example.—What **I** beam is required to span an opening of 30 feet, to be trussed 3 feet deep between centres in the manner illustrated in Fig. 6, page 165? (this trussed beam carries a brick wall which weighs 500 lbs. per lineal foot, but which braces the beam from yielding sideways), the beam to be proportioned for a safety factor of four ?

Here the beam can be considered as composed of two separate beams, reaching from the centre to each end, each being 15 feet long, carrying a distributed load of $15 \times 500 = 7,500$ lbs., and subject to a compression resulting from the trussing of 18,750 lbs. Our approximate tables for beams, on page 69, will be found most convenient for such calculations as the above, and are sufficiently accurate for practical purposes. For **I** beam, dividing co-efficient by 4 we have $\dfrac{1.05 \, A \, D}{L} =$ safe distributed load $= 3.75$ tons.

By trial we find for an 8″ 65 lb. **I** beam $\dfrac{1.05 \times 6.5 \times 8}{15} = 3.64$, or nearly correct.

For the compression, allowing 12,500 lbs. per square inch, we require $1\frac{1}{2}$ square inches. Therefore an 8″ **I** beam, 8 square inches section, will be safe.

If desirable to use a deeper, lighter beam, try a 9-inch beam 75 lbs. per yard ; allowing $1\frac{1}{2}$ square inches for the compression, we have a section of 6 square inches remaining ; $\dfrac{105 \times 6 \times 9}{15} = 3.78$. The latter beam being both stronger and lighter than the 8-inch.

(2.) When the beam is subject to compression and is liable to fail like a horizontal strut by lateral flexure.

Rule.—Consider first the resistance as a strut and then make the necessary increment of section to resist the bending stress, remembering that if the addition is made to the flanges then only flange stresses have to be considered, but if the increased area is obtained by thickening the web of I beam or channel sections, then the additional area so obtained should be treated as a rectangular section whose thickness is the amount added to the web, and whose depth is the depth of the beam.

Example.—A trussed girder of the form exhibited in Fig. 8, page 165, is a box section made up of two channels separated with flanges outward, and plated top and bottom. The whole girder is 30 feet long and is loaded 1,000 lbs. per lineal foot. The compression resulting from the trussing is 25,000 lbs. The structure has no lateral bracing. What will be safe proportions for it, the stresses not to exceed $\frac{1}{5}$ of the ultimate ?

It is evident that we have to consider it as a flat-ended strut 30 feet long liable to fail horizontally, and also as a series of 3 beams each 10 feet long and loaded with 10,000 lbs. evenly distributed. Trying 2 lightest 5" channels, each 2.27 square inches section, separated $5\frac{1}{4}$" so as to be covered by 9" plates, we have (omitting the plates in this calculation,) the radius of gyration around vertical axis (see page 110) = 3.25 inches, $\frac{l}{r} = 110$, one-fifth of ultimate (by Table I, page 118) = 5,600 lbs. per square inch, or 5,600 × $4\frac{1}{2}$ = 25,200 lbs. safe resistance, which is ample. Now proportioning the plates to resist the bending strain we have maximum bending moments (see page 78), $\dfrac{120 \times 10,000}{8} = 150,000$ inch-lbs.

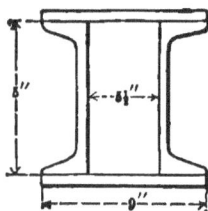

The plates act with a leverage equal to the depth of the channel, viz., 5"; $\dfrac{150,000}{5} = 30,000$ lbs. tension on top or compression on bottom plate, which, allowing for 10,000 lbs. per square inch, and allowing for loss by rivets, will require a plate $\frac{3}{8}$" thick.

(3) Taking the last example, if it was desired to form the section out of a pair of channels latticed top and bottom with no cover plates, we would have to consider the section added to the channels (being on the web alone), as a simple rectangular section. By the formula on page 69, approximate rules, we find that such a section only 5" deep would require a thickness of 3.8 inches, which is impracticable; we have therefore to use deep-

er and heavier channels. Trying 8″ channels separated as before 5½ inches, with flanges outward, and having radius of gyration for the pair around vertical axis = 3.4, $\dfrac{l}{r}$ = 106. Safe load

$\dfrac{29,000}{5}$ = 5,800 lbs. per square inch. As the compression is 25,000

lbs., there is required 4.3 square inches for this purpose. By

formula 2, page 68, $\dfrac{.52 \times \text{area} \times 8}{10}$ = 5 tons, from which is

found the area required to resist bending = 12 square inches. 12 + 4.3 = 16.3 square inches for 2 channels, or the heaviest 8″ channels 80 lbs. per yard would be required.

By the same method we find 10′ channels 68 lbs. per yard, will answer the purpose, or our lightest 12″ channels 60 lbs. per yard, will exactly meet the requirements and be the lightest channel that can be used in the manner proposed for the purpose.

In cases where the load is concentrated at the truss points, there being no bending stress, the resistance as a strut has only to be considered, and when braced laterally the strut length is reduced to the distances between bracing.

ELEMENTS OF PENCOYD STRUCTURAL SHAPES.

In the following tables, pages 88, 91, various properties of rolled structural iron are given, whereby the strength or stiffness of any shape can be readily determined.

SYMBOLS.

$I =$ Moment of inertia.
$E =$ Modulus of elasticity.
$W =$ Load on beam in net tons.
$w =$ Load on beam in pounds.
$R =$ Radius of gyration.
$A =$ Total area of cross section.
$L =$ Length between supports in feet.
$l =$ Length between supports in inches.

Column I.—Chart number.

Columns III. to VI.—Details of the sectional areas in square inches. The flanges being taken the entire width of section, and the web considered between the flanges.

Columns VII. and VIII.—The moments of inertia, respectively, at right angles to and parallel with web of beam. In all cases the axes referred to pass through the centre of gravity of the cross-section, as illustrated at the head of each table.

Columns IX. and X.—The radii of gyration in inches $= \sqrt{\dfrac{I}{A}}$.

When R^2 is required, simply divide the moment of inertia by the area of the section. The values of I and R have all been carefully calculated by the formulæ given on pages 102–111. The tables give the value of I for the minimum section of each particular shape, but the section can be increased in area up to the maximum limit given in the descriptive tables, pages 2–12, and the value of I can be readily obtained for any enlarged section as described on pages 106–108.

Column XI.—Co-efficient for the greatest safe load evenly distributed over the beam. This is the calculated load in net tons for a beam of the given size and section, one foot long, and is derived from the formula $\dfrac{Wl}{8} = \dfrac{7I}{\frac{1}{2}\ \text{depth of beam}}$, which gives results averaging one-third of the ultimate strength of the beam. The safe distributed load for any beam of the size and section given in Columns II. to VI. can be found by dividing the corresponding co-efficient in column XI. by the length of the beam between supports, in feet.

Example.—The greatest safe load that can be evenly distributed on a beam 10 inches deep having a sectional area of 9.04 square inches and spanning 12 feet is $\dfrac{138.4}{12} = 11.5$ tons.

If the load is concentrated in the middle of the beam, one-half this result, or 5.75 tons, is the greatest safe load.

If the sectional area of the beam is increased, find the moment of inertia for the increased section as described on page 106, and

the co-efficient for a distributed safe load $= \dfrac{9\frac{1}{3}I}{\text{depth of beam}}$.

Example.—The 10" beam taken in last example, 9.04 square inches area, is increased to 10.6 square inches section. The inertia of enlarged section is found as per formulæ on page 106,

$$\frac{1.56 = (\text{increase of area}) \times 100 = (\text{square of depth})}{12} = 13. + 148.3$$

(inertia, col. vii., page 92,) $= 161.3$ or moment of inertia desired.

Co-efficient for safe load $= \dfrac{161.3 \times 9\frac{1}{3}}{10} = 150.5$. Dividing this

co-efficient by the span in feet (12), gives $\dfrac{150.5}{12} = 12.54$ tons as the maximum safe load distributed, or 6.27 tons in the middle of the beam.

Lateral Flexure.—It will be noted that when subjected to such loads as above obtained, the beams are presumed to be secured from bending sideways, and it will be safest to limit the application to beams secured laterally at intervals, in length not exceeding twenty times the width of flange. See preface to tables of safe loads for beams, page 36.

Columns XII. and XIII.—Deflections.

The figures in the tables are the calculated deflections for beams of the sizes and sections given, one foot long between bearings and supporting a load of one ton. They are derived by

means of the formulæ $\dfrac{wl^3}{48EI} =$ deflection for load in middle of

beam. $\dfrac{wl^3}{76.8EI} =$ deflection for load evenly distributed.

The modulus of transverse elasticity is assumed as 26,000,000 lbs. The elasticity of rolled iron is somewhat uncertain, it is frequently quoted as high as 29,000,000 lbs., and experiments on bars of exceptionally stiff iron will often give results much in excess of this. But recent experiments on rolled beams show that 26,000,000 lbs. is a fair average for this form of wrought iron. See page 19.

The deflection of any beam of the sectional area given in cols. IV. to VI., and loaded within the elastic limit, is found by multiplying the corresponding co-efficient in cols. XII., XIII., by the weight in tons and the cube of the length in feet.

Example.—A 12″ I beam, 11.95 square inches section, 13 feet between supports, carries an evenly distributed load of 15 tons. Deflection = .0000063 × 15 × 13³ = .207 inches.

If the sectional area of this shape is increased, the value of I for the enlarged section must be found as described in previous example. By reducing the formulæ for deflection to their simplest forms we obtain :

$$\frac{WL^3}{362I} = \text{deflection in inches for load in middle.}$$

$$\frac{WL^3}{580I} = \text{deflection in inches for distributed load.}$$

Example.—The 12″ beam in previous example 11.95 square inches area, is increased to 13.8 square inches. The inertia of enlarged section is found as per formula, page 106.

$$\frac{1.85\,(\text{increase of area}) \times 144\,(\text{square of depth})}{12} = 22.2 + 272.86$$

inertia, col. vii., page 92, = 295.06, or moment of inertia desired.

$$\text{Deflection} = \frac{15 \times 13^3}{580 \times 295.06} = .19 \text{ inches.}$$

For beams of the same depth, but of any sectional area, the deflection remains uniform so long as the loads bear a uniform ratio to the strength of the beam. For this reason, the single column of deflections applies to any section of the same size of beam, in the tables of safe loads.

Column XIV.—Maximum load in tons.

There is a limit in the length of beams at which the rule for safe loading ceases to apply. This point is reached when the load attains the safe limit of resistance offered by the web of the beam against crippling.

The maximum load can be placed on any beam shorter than the length indicated, but must not be exceeded. It is obtained by Gordon's formula, taking 6 tons per square inch as the safe resistance of wrought iron to crushing.

$$W = \frac{6dt}{1 + \dfrac{l^2}{3000t^2}}$$

d = depth of beam.

t = thickness of web.

l = d × secant 45° ($l^2 = 2d^2$).

Example.—An 8'' 65 lb. beam has a maximum load of 10.46 tons, which corresponds to the greatest safe load on a beam of this section, 7.7 feet between supports, if the load is distributed, or 3.85 feet if the load is at middle of beam. If this shape is increased to 7½ square inches area, having a web $\frac{7}{16}$'' thick, then maximum safe load becomes

$$W = \frac{6'' \times 8'' \times \frac{7}{16}''}{1 + \dfrac{128}{(3000 \times \frac{7}{16}{}^2)}} = 17.2 \text{ tons.}$$

ELEMENTS OF PENCOYD BEAMS.

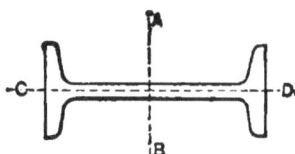

I.	II.	IIL	IV.	V.	VI.	VII.	VIII.
			AREAS IN SQUARE INS.			MOMENT OF INERTIA.	
CHART NUM- BER.	SIZE IN INCHES.	WEIG'T PER YARD.	Flanges	Web.	Total.	Axis A. B.	Axis C. D.
1	15	200	11.86	8.04	19.90	682.08	28.50
2	15	145	8.97	5.58	14.55	521.19	16.91
3	12	168	10.66	6.23	16.89	371.98	23.19
4	12	120	7.42	4.53	11.95	272.86	12.22
5	10½	134	9.57	3.87	13.44	241.63	19.00
5½	10⅓	108	7.33	3.50	10.83	195.42	12.45
6	10½	89	5.91	3.03	8.94	162.26	8.34
7	10	112	7.23	3.94	11.17	173.58	10.64
8	10	90	6.29	2.75	9.04	148.31	8.09
9	9	90	6.15	2.92	9.07	118.81	8.44
10	9	70	4.77	2.21	6.98	94.44	5.59
11	8	81	5.58	2.56	8.14	83.93	7.23
12	8	65	4.50	2.03	6.53	69.17	5.02
13	7	65	4.17	2.41	6.58	49.78	4.15
14	7	52	3.84	1.30	5.14	43.08	3.43
15	6	50	3.16	1.88	5.04	26.92	2.15
16	6	40	2.91	1.17	4.08	24.10	1.80
17	5	34	2.13	1.25	3.38	13.40	1.21
18	5	30	2.06	.88	2.94	12.50	1.09
19	4	28	2.15	.75	2.90	7.69	1.17
20	4	18.5	1.34	.56	1.90	5.14	.49
21	3	23	1.72	.53	2.25	3.29	.77
22	3	17	1.37	.34	1.71	2.66	.48

ELEMENTS OF PENCOYD BEAMS.

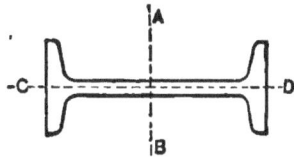

IX.	X.	XI.	XII.	XIII.	XIV.	II.	I.
RADII OF GYRATION.		CO-EFFICIENT SAFE LOAD DISTRIBUTED.	CO-EFFICIENT FOR DEFLECTION.		MAXIMUM LOAD IN TONS.	SIZE IN INCHES.	CHART NUMBER.
Axis A. B.	Axis C. D.		Load in Centre.	Load Distributed.			
5.86	1.20	424.41	.0000041	.0000025	43.20	15	1
5.98	1.08	324.30	.0000053	.0000033	22.10	15	2
4.69	1.17	289.32	.0000074	.0000046	38.63	12	3
4.78	1.01	212.22	.0000101	.0000063	22.22	12	4
4.24	1.19	214.78	.0000115	.0000072	22.13	10¼	5
4.25	1.07	173.71	.0000142	.0000089	17.71	10½	5½
4.26	.97	144.23	.0000171	.0000107	13.85	10½	6
3.94	.98	162.02	.0000159	.0000099	23.68	10	7
4.05	.95	138.43	.0000186	.0000116	13.18	10	8
3.62	.96	123.21	.0000232	.0000145	16.53	9	9
3.68	.89	97.94	.0000292	.0000183	9.94	9	10
3.21	.94	97.92	.0000329	.0000205	15.49	8	11
3.25	.88	80.70	.0000399	.0000249	10.46	8	12
2.75	.79	66.38	.0000546	.0000341	15.69	7	13
2.89	.82	57.44	.0000640	.0000400	6.17	7	14
2.31	.65	41.87	.0001025	.0000641	12.77	6	15
2.43	.66	37.49	.0001144	.0000715	6.50	6	16
1.99	.60	25.01	.0002059	.0001287	8.01	5	17
2.06	.60	23.33	.0002206	.0001379	4.86	5	18
1.63	.63	17.94	.0003589	.0002243	5.12	4	19
1.65	.51	12.00	.0005366	.0003354	3.03	4	20
1.21	.59	10.24	.0008382	.0005239	4.11	3	21
1.25	.53	8.28	.0010366	.0006479	2.34	3	22

ELEMENTS OF PENCOYD CHANNELS.

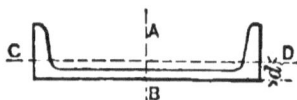

I.	II.	III.	IV.	V.	VI.	VII.	VIII.
CHART NUM- BER.	SIZE IN INCHES.	WEIG'T PER YARD.	AREAS IN SQUARE INS.			MOMENTS OF INERTIA	
			Flanges	Web.	Total.	Axis A. B.	Axis C. D.
30	15	148	6.50	8.36	14.86	451.51	19.05
31	12	88.5	4.59	4.24	8.83	182.71	7.42
32	12	60	2.87	3.07	5.94	123.71	3.22
34	10	60	3 56	2.43	5.99	92.08	4.20
35	10	49	2.67	2.22	4.89	73.91	2.33
36	9	54	2.97	2.43	5.40	64.34	2.47
37	9	37	1.81	1.91	3.72	43.65	1.81
38	8	43	2.28	1.97	4.25	40.00	2.17
39	8	30	1.34	1.62	2.96	28.23	1.06
40	7	41	2.30	1.80	4.10	29.51	1.71
41	7	26	1.38	1.26	2.64	18.46	.90
42	6	33	2.04	1.25	3.29	18.37	1.46
44	6	23	1.09	1.18	2.27	11.67	.59
45	5	27.3	1.69	1.04	2.73	10.29	.86
46	5	19	.91	.97	1.88	6.67	.37
47	4	21.5	1.34	.81	2.15	5.16	.54
48	4	17.5	1.02	.73	1.75	4.14	.41
49	3	15	.86	.66	1.52	2.03	.32
50	2¼	11.3	.69	.44	1.13	.80	.21
51	2	8.75	.55	.33	.88	.48	.08

ELEMENTS OF PENCOYD CHANNELS.

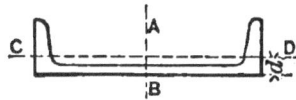

IX.	X.	XI.	XII.	XIII.	XIV.	XV.	II.	I.
RADII OF GYRATION.		CO-EFFICIENT SAFE LOAD DISTRIBUTED.	CO-EFFICIENTS FOR DEFLECTION.		MAXIMUM LOAD IN TONS.	DISTANCE, d, FROM BASE TO NEUTRAL AXIS.	SIZE IN INCHES.	CHART NUMBER.
Axis A. B.	Axis C. D.		Load in centre.	Load distributed.				
5.51	1.13	280.94	.0000061	.0000038	40.64	.95	15	30
4.55	.92	142.11	.0000151	.0000094	18.49	.71	12	31
4.56	.74	96.22	.0000223	.0000139	9.14	.62	12	32
3.92	.84	85.94	.0000298	.0000186	9.10	.75	10	34
3.89	.69	68.98	.0000374	.0000234	7.25	.64	10	35
3.45	.68	66.73	.0000429	.0000268	10.87	.67	9	36
3.43	.59	45.27	.0000632	.0000395	6.38	.55	9	37
3.06	.71	46.66	.0000690	.0000431	8.77	.60	8	38
3.09	.60	32.94	.0000977	.0000611	4.79	.50	8	39
2.68	.65	39.35	.0000935	.0000584	9.07	.65	7	40
2.64	.58	24.61	.0001495	.0000934	3.42	.48	7	41
2.36	.67	28.58	.0001501	.0000938	6.50	.66	6	42
2.27	.51	18.16	.0002363	.0001477	5.24	.46	6	44
1.93	.56	19.21	.0002680	.0001675	5.92	.61	5	45
1.88	.45	12.45	.0004136	.0002585	4.86	.42	5	46
1.55	.50	12.03	.0005349	.0003343	5.12	.53	4	47
1.54	.48	9.65	.0006667	.0004167	4.29	.45	4	48
1.16	.46	6.32	.0013584	.0008490	3.49	.51	3	49
.85	.43	3.33	.0034350	.0021470	3.20	.46	2¼	50
.74	.31	2.25	.0057230	.0035770	2.49	.37	2	51

ELEMENTS OF PENCOYD DECK BEAMS.

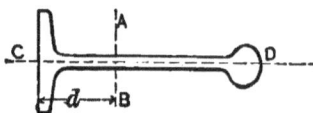

I.	II.	III.	IV.	V.	VI.	VII.	VIII.
Chart Num-ber.	Size in Inches.	Weig't per Yard.	Areas in Square Ins.			Moments of Inertia	
			Fl'ge	Bulb.	Web.	Total.	Axis A. B.	Axis C. D.
60	12	104	3.59	2.89	3.90	10.38	221.98	9.33
61	11	91	3.26	2.52	3.28	9.06	164.09	7.64
62	10	80	2.87	2.19	2.96	8.02	118.22	6.13
63	9	72	2.50	2.06	2.61	7.17	84.77	4.92
64	8	61	2.17	1.85	2.09	6.11	57.66	3.63
65	7	52	1.86	1.55	1.80	5.21	34.40	2.59
66	6	42	1.52	1.28	1.38	4.18	21.95	1.64
67	5	34	1.22	1.04	1.11	3.37	12.04	.98

ELEMENTS OF PENCOYD DECK BEAMS.

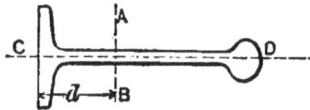

IX.	X.	XI.	XII.	XIII.	XIV.	XV.	II.	I.
RADII OF GYRATION.		CO-EFFICIENT SAFE LOAD DISTRIBUTED.	CO-EFFICIENTS FOR DEFLECTION.		MAXIMUM LOAD IN TONS.	DISTANCE, d, FROM BASE TO NEUTRAL AXIS.	SIZE IN INCHES.	CHART NUMBER.
Axis A. B.	Axis C. D.		Load in centre.	Load distributed.				
4.62	.95	172.6	.0000122	.0000078	18.50	5.24	12	60
4.25	.92	139.5	.0000168	.0000105	15.72	4.68	11	61
3.84	.87	110.3	.0000233	.0000146	15.26	4.27	10	62
3.44	.83	87.9	.0000325	.0000203	14.63	4.00	9	63
3.07	.77	67.3	.0000478	.0000299	12.12	3.50	8	64
2.57	.71	45.8	.0000802	.0000501	11.30	3.20	7	65
2.29	.63	34.2	.0001257	.0000785	9.03	2.65	6	66
1.89	.54	22.4	.0002291	.0001432	8.01	2.22	5	67

7

ELEMENTS OF PENCOYD ANGLES.

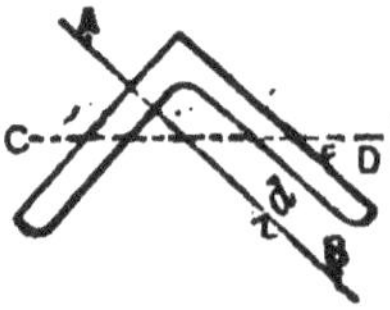

EVEN LEGS.

I.	II.	III.	IV.	V.	VI.	VII.	VIII.
Chart Number	Size in Inches.	Weight per Yard.	Moments of Inertia.		Radii of Gyration.		Distance, *d*, from Base to Neutral Axis.
			Axis A. B.	Axis C. D.	Axis A. B.	Axis C. D.	
120	6 × 6 × 7/16	50.6	17.68	7.16	1.87	1.19	1.66
	6 × 6 × 1	110.0	35.46	15.00	1.80	1.17	1.86
121	5 × 5 × 7/16	41.8	10.02	4.16	1.55	1.00	1.41
	5 × 5 × 1	90.0	19.64	8.67	1.48	.98	1.61
122	4 × 4 × 3/8	28.6	4.36	1.86	1.24	.81	1.14
	4 × 4 × 3/4	54.4	7.67	3.45	1.19	.80	1.27
123	3½ × 3½ × 3/8	24.8	2.87	1.20	1.07	.70	1.01
	3½ × 3½ × 5/8	39.8	4.33	1.85	1.04	.69	1.10
124	3 × 3 × 1/4	14.4	1.24	.51	.93	.60	.84
	3 × 3 × 5/8	33.6	2.62	1.15	.88	.59	.98
125	2¾ × 2¾ × 1/4	13.1	.95	.39	.85	.55	.78
	2¾ × 2¾ × 1/2	25.0	1.67	.72	.82	.54	.87
126	2½ × 2½ × 1/4	11.9	.70	.29	.77	.50	.72
	2½ × 2½ × 1/2	22.5	1.23	.54	.74	.49	.81
127	2¼ × 2¼ × 1/4	10.6	.50	.21	.69	.45	.65
	2¼ × 2¼ × 7/16	17.8	.79	.34	.67	.44	.72
128	2 × 2 × 3/16	7.1	.27	.11	.62	.40	.57
	2 × 2 × 3/8	13.6	.50	.21	.61	.39	.64
129	1¾ × 1¾ × 3/16	6.2	.18	.08	.53	.36	.51
	1¾ × 1¾ × 3/8	11.7	.31	.14	.51	.35	.57
130	1½ × 1½ × 3/16	5.3	.11	.05	.46	.31	.44
	1½ × 1½ × 3/8	9.8	.19	.09	.44	.31	.51
131	1¼ × 1¼ × 1/8	3.0	.05	.02	.41	.26	.36
	1¼ × 1¼ × 1/4	5.6	.08	.04	.38	.26	.40
132	1 × 1 × 1/8	2.3	.02	.01	.29	.20	.30
	1 × 1 × 1/4	4.4	.04	.02	.29	.20	.35

ELEMENTS OF PENCOYD ANGLES.

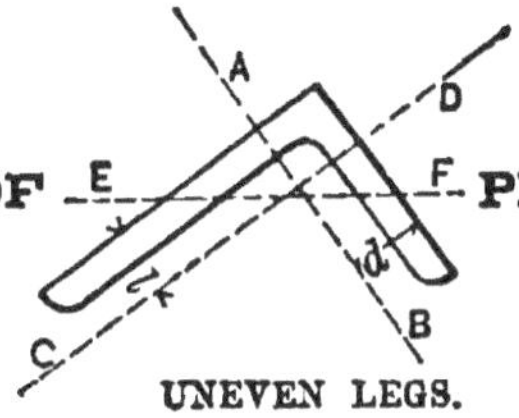

UNEVEN LEGS.

I.	II.	III.	IV.	V.	VI.	VII.	VIII.	IX.	X.	XI.
Chart Number	Size in Inches.	Weight per Yard.	Mom. of Inertia.			Radii of Gyration.			Dist. from Base to Neut. Axes	
			Axis A. B.	Axis C. D.	Axis E. F.	Axis A. B.	Axis C. D.	Axis E. F.	*d.*	*l.*
140	6 × 4 × 7/16	41.8	15.46	5.60	3.55	1.92	1.16	.92	1.96	.96
	6 × 4 × 1	90.0	30.75	10.75	7.46	1.85	1.09	.91	2.17	1.17
141	5 × 4 × 3/8	32.3	8.14	4.66	2.47	1.59	1.20	.87	1.53	1.03
	5 × 4 × 1	80.0	18.17	10.17	6.10	1.51	1.13	.86	1.75	1.25
142	5 × 3½ × 3/8	30.5	7.78	3.23	1.95	1.60	1.03	.80	1.61	.86
	5 × 3½ × 3/4	58.1	13.92	5.55	3.72	1.55	.98	.79	1.75	1.00
143	5 × 3 × 3/8	28.6	7.37	2.04	1.42	1.61	.85	.70	1.70	.70
	5 × 3 × 3/4	54.4	13.15	3.51	2.58	1.55	.80	.69	1.84	.84
144	4½ × 3 × 3/8	26.7	5.50	1.98	1.27	1.44	.86	.69	1.49	.74
	4½ × 3 × 5/8	43.0	8.44	2.98	2.04	1.40	.83	.68	1.58	.83
145	4 × 3½ × 3/8	26.7	4.17	2.99	1.44	1.25	1.06	.74	1.20	.95
	4 × 3½ × 5/8	43.0	6.37	4.52	2.34	1.22	1.03	.73	1.29	1.04
146	4 × 3 × 3/8	24.8	3.96	1.92	1.10	1.26	.88	.67	1.28	.78
	4 × 3 × 5/8	39.8	6.03	2.87	1.69	1.23	.85	.65	1.37	.87
147	3½ × 3 × 11/32	21.2	2.53	1.72	.86	1.09	.90	.64	1.07	.82
	3½ × 3 × 5/8	36.7	4.11	2.81	1.49	1.06	.87	.64	1.17	.92
148	3 × 2½ × 5/16	16.2	1.42	.90	.47	.94	.74	.54	.93	.68
	3 × 2½ × 1/2	25.0	2.08	1.30	.72	.91	.72	.54	1.00	.75
149	3 × 2 × 1/4	11.9	1.09	.39	.25	.96	.58	.46	.99	.49
	3 × 2 × 1/2	22.5	1.92	.67	.47	.92	.55	.46	1.08	.58
150	3½ × 2½ × 5/16	17.8	2.19	.94	.56	1.11	.73	.56	1.14	.64
	3½ × 2½ × 1/2	27.5	3.24	1.36	.87	1.08	.70	.56	1.20	.70
151	6 × 3½ × 7/16	39.6	14.76	3.81	2.68	1.93	.98	.82	2.06	.81
	6 × 3½ × 1	85.0	29.24	7.21	5.75	1.86	.92	.81	2.26	1.01
152	6½ × 4 × 7/16	44.0	19.29	5.72	3.87	2.09	1.14	.94	2.18	.93
	6½ × 4 × 1	95.0	38.66	11.00	8.35	2.02	1.08	.93	2.38	1.13
153	5½ × 3½ × 3/8	32.3	10.12	3.27	2.14	1.77	1.05	.81	1.82	.82
	5½ × 3½ × 5/8	52.3	15.73	4.96	3.35	1.73	.97	.80	1.91	.91
154	7 × 3½ × 5/8	61.7	30.25	5.28	4.45	2.21	.92	.85	2.57	.82
	7 × 3½ × 1	95.0	45.37	7.53	6.70	2.19	.88	.84	2.71	.96
155	2½ × 2 × 1/4	10.6	.71	.37	.20	.81	.59	.43	.78	.54
	2½ × 2 × 1/2	20.0	1.09	.63	.37	.74	.56	.43	.87	.62
156	2¼ × 1½ × 3/16	6.7	.34	.13	.08	.71	.43	.34	.76	.38
	2¼ × 1½ × 3/8	12.6	.50	.21	.15	.63	.40	.34	.82	.44
157	2 × 1¼ × 3/16	5.7	.23	.07	.05	.63	.35	.31	.68	.31
	2 × 1¼ × 3/8	9.2	.32	.10	.08	.59	.33	.29	.70	.32

ELEMENTS OF PENCOYD TEES.

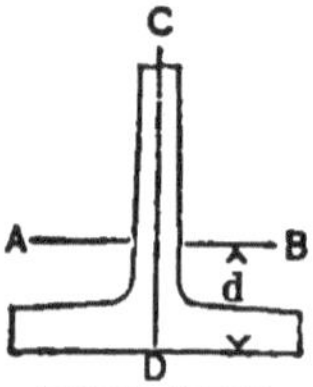

EVEN LEGS.

I.	II.	III.	IV.	V.	VI.	VII.	VIII.
Chart Number	Size in Inches.	Weight per Yard.	Moments of Inertia.		Radii of Gyration.		Distance, d, from Base to Neutral Axis.
			Axis A. B.	Axis C. D.	Axis A. B.	Axis C. D.	
70	4 × 4 × ½	36.5	5.26	2.55	1.20	.84	1.14
71	3½ × 3½ × $\frac{15}{32}$	31.	3.47	1.70	1.06	.74	1.00
72	3 × 3 × $\frac{15}{32}$	26.	2.10	1.01	.90	.62	.90
73	2½ × 2½ × $\frac{7}{16}$	19.5	1.12	.58	.78	.55	.75
74	2½ × 2½ × ⅜	17.52	.97	.49	.75	.53	.75
75	2¼ × 2¼ × ¼	11.75	.52	.30	.65	.50	.61
76	2¼ × 2¼ × $\frac{9}{32}$	12.	.54	.27	.67	.47	.65
77	2 × 2 × $\frac{9}{32}$	10.5	.38	.19	.60	.43	.60
78	1¾ × 1¾ × $\frac{7}{32}$	7.1	.21	.10	.54	.37	.50
79	1½ × 1½ × $\frac{7}{32}$	6.	.13	.06	.46	.32	.45
80	1¼ × 1¼ × $\frac{3}{16}$	4.5	.07	.04	.37	.27	.37
81	1 × 1 × $\frac{3}{16}$	3.0	.03	.02	.30	.26	.30
82	3 × 3 × $\frac{11}{32}$	19.3	1.59	.75	.91	.62	.84
83	3 × 3 × $\frac{13}{32}$	22.6	1.83	.89	.90	.63	.86

ELEMENTS OF^A PENCOYD TEES.

UNEVEN LEGS.

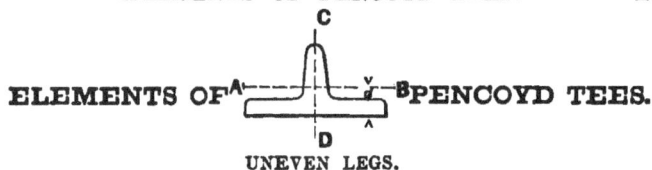

I.	II.	III.	IV.	V.	VI.	VII.	VIII.
Chart Number	Size in Inches.	Weight per Yard.	Moments of Inertia.		Radii of Gyration.		Distance, d, from Base to Neutral Axis.
			Axis A. B.	Axis C. D.	Axis A. B.	Axis C. D.	
90	4½ × 3½	44.5	5.27	3.66	1.09	.91	1.16
91	4 × 3½	41.8	4.65	3.23	1.05	.88	1.09
92	5 × 2½	30.7	1.61	4.01	.72	1.14	.67
93	5 × 2½	33.0	1.63	4.58	.70	1.17	.64
94	4 × 3	25.9	1.94	2.18	.86	.92	.77
95	4 × 3	25.25	2.09	1.69	.91	.82	.84
96	4 × 2	20.4	.68	1 68	.58	.91	.54
97	3 × 3½	28.25	3.12	1.06	1.05	.61	1.10
98	3 × 2½	23.8	1.38	.94	.76	.63	.82
99	3 × 1½	11.2	.19	.56	.41	.71	.37
100	2½ × 1¼	9.1	.10	.33	.33	.60	.32
101	2 × 1½	8.75	.16	.18	.43	.45	43
102	2 × 1	7.	.05	.17	.26	.49	.27
103	2 × ⁹⁄₁₆	5.88	.01	.17	.13	.54	.17
104	2¼ × 1¾	18.75	.56	.62	.55	.58	.66
105	2¾ × 2	21.	.83	.63	.63	.55	.75
106	5 × 3½	48.44	5.37	5.31	1.05	1.04	1.05
107	5 × 4	44.1	6.24	5.25	1.19	1.09	1.08
108	2¼ × ⁹⁄₁₆	6.5	.01	.24	.12	.61	.18
109	4 × 4½	38.5	7.26	2.70	1.37	.84	1.32
110	3 × 2½	17.6	.94	.74	.73	.65	.69
111	3 × 2½	20.6	1.08	.89	.72	.66	.70

MOMENTS OF INERTIA.

The following formulæ were used in calculating the moments of inertia and radii of gyration of the various sections given in the tables, pages 92–101.

When not otherwise specified the axis referred to passes through the centre of gravity of the section, in a horizontal position to the figure as shown.

I signifies moment of inertia.
A " total area of section.
R " radius of gyration.
d " distance from base to centre of gravity.

In all cases the radius of gyration $= \sqrt{\dfrac{I}{A}}$, and the moment of

$$\text{resistance} = \frac{I \times \text{co-efficient for strength of material}}{\text{distance from neutral axis to farthest edge of section}}.$$

SOLID RECTANGLE.

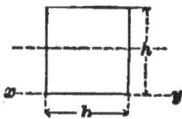

$$I = \frac{bh^3}{12} = \frac{Ah^2}{12}.$$

$$I, \text{ axis } xy = \frac{bh^3}{3}.$$

HOLLOW RECTANGLE OR I BEAM WITH PARALLEL FLANGES.

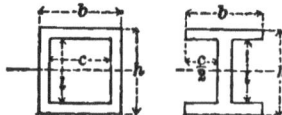

$$I = \frac{bh^3 - cl^3}{12}.$$

SOLID TRIANGLE.

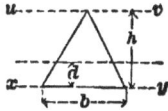

$$I = \frac{bh^3}{36}.$$

$$I, \text{axis } xy = \frac{bh^3}{12}.$$

$$I, \text{axis } uv = \frac{bh^3}{3}.$$

$$d = \frac{h}{3}.$$

SOLID CIRCLE.

$$I = .7854 \, r^4 = \frac{AD^2}{16}.$$

HOLLOW CIRCLE.

$$I = (\text{outer radius}^4 - \text{inner radius}^4) \, .7854.$$

SOLID SEMICIRCLE.

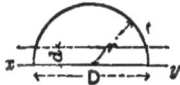

$$I = .11r^4.$$

$$I, \text{axis } xy = .3927r^4 = \frac{AD^2}{16}.$$

$$d = .4244r.$$

SOLID ELLIPSE.

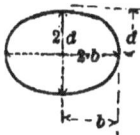

$$I = .7854bd^3.$$

TEE SECTION.

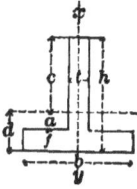

$$I = \frac{tc^3 + bd^3 - (b - t)\,a^3}{3}.$$

$$I,\ \text{axis } xy = \frac{fb^3 + (h - f)\,t^3}{12}.$$

$$d = \frac{\dfrac{bf^2}{2} + (h - f)\,t\left(f + \dfrac{h - f}{2}\right)}{A} = \frac{bf^2 + t\,(h^2 - f^2)}{2A}.$$

ANGLE SECTION.

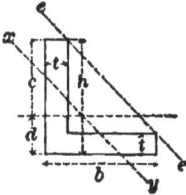

$$I = \frac{tc^3 + bd^3 - (b - t)\,(d - t)^3}{3}.\quad \text{For even or}$$

uneven angles.

$$I,\ \text{axis } uv = \frac{t\,(b - d_1)^3 + hd_1{}^3 - (h - t)(d_1 - t)^3}{3}.$$

For uneven angles.

xy passes through centre of gravity parallel to ee.

$$I \text{ axis } xy = \frac{2d^4 - 2(d - t)^4 + t\left[b - \left(2d - \dfrac{t}{2}\right)\right]^3}{3}.\quad \text{For even}$$

angles.

A close approximation for the latter is the following :

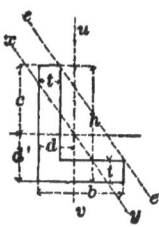

$$I,\ \text{axis } xy = \frac{Ab^2}{25}.\quad \text{For even angles.}$$

$$I,\ \text{axis } xy = \frac{Ah^2b^2}{13\,(h^2 + b^2)}\quad \text{For uneven angles.}$$

$$d = \frac{bt^2 + t\,(h^2 - t^2)}{2A}.\quad \text{For even and uneven an-}$$

gles.

$$d' = \frac{ht^2 + t\,(b^2 - t^2)}{2A}.\quad \text{For uneven angles.}$$

In even angles radius of gyration around $xy = $ two-thirds ($\frac{2}{3}$) of the radius of gyration around horizontal axis.

In uneven angles the distance from centre of gravity in direction of the long leg exceeds that in the direction of the short leg by half the difference in the length of the two legs.

I BEAM SECTION.

$s = $ taper of flange.

$$l = k - \frac{2s}{3}.$$

$$I = \frac{bh^3 - ck^3}{12} + \frac{cs^3}{18} + \frac{csl^2}{4}.$$

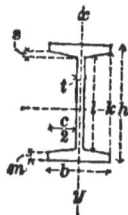

$$I, \text{ axis } x\,y = \frac{mb^3}{6} + \frac{kt^3}{12} + \frac{s\left(\frac{b-t}{2}\right)^3}{9} +$$

$$2\,s\left(\frac{b-t}{2}\right)\left(\frac{b}{6} + \frac{t}{3}\right)^2.$$

CHANNEL SECTION.

$s = $ taper of flange.

$$r = \frac{s}{b - t}.$$

$$I = \frac{bh^3 - \dfrac{1}{8\,r}\left(k^4 - l^4\right)}{12}.$$

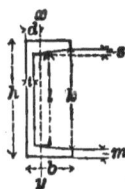

$$I, \text{ axis } xy = \frac{2\,mb^3 + lt^3 + \dfrac{r}{2}\left(b^4 - t^4\right)}{3} - Ad^2.$$

$$d = \frac{mb^2 + \dfrac{kt^2}{2} + \dfrac{s}{3}\left(b - t\right)\left(b + 2t\right)}{A}.$$

DECK BEAM SECTION.

$s =$ taper of flange. $\qquad a =$ area of bulb.

$$o = m - \frac{s}{3}.$$

$$I = \frac{aw^2}{15} + al^2 + \frac{tc^3}{3} + \frac{bd^3}{3} - \frac{m^3 (b - t)}{3} +$$

$$\frac{(b - t)s^3}{36} + \frac{s (h - t) o^2}{2}.$$

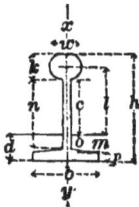

$$I, \text{ axis } xy = \frac{ak^2}{12.4} + \frac{nt^3}{12} + \frac{\left(p + \frac{s}{4}\right)b^3}{12}$$

$$d = \frac{a (2 h - k) + t (h - k)^2 + (b - t) p^2 + s(b - t) \left(p + \frac{s}{3}\right)}{2A}.$$

In the table of elements, pages 92–101, the moments of inertia and radii of gyration are given for the minimum section of each shape. but the moment of inertia.for any increased section can readily be ascertained as follows, without recalculating the whole.

FOR ANY I BEAM, CHANNEL BAR OR DECK BEAM.

AXIS PERPENDICULAR TO WEB.

Let $a =$ increase of area in square inches over minimum section given in the table. Let $d =$ depth (size) of beam, then $\frac{ad^2}{12}$ is the moment of inertia for increase of area, which added to tabular figures gives the correct result for the enlarged section.

Example.—A 12″ I Beam, No. 4, area 12 square inches, is increased to 14 square inches. $\frac{2 \times 12^2}{12} = 24$, which added to the moment given in col. 7—272.86 + 24 = 296.86, the moment of inertia desired.

Radius of gyration of the former $\sqrt{\dfrac{272.86}{12}} = 4.78$ inches.

Radius of gyration of the latter $\sqrt{\dfrac{296.86}{14}} = 4.60$ inches.

The radius of gyration will be found to alter very little, and for all practical purposes, the tabular figures may be accepted within the range of section possible for each shape.

The above is only a close approximation for deck beams.

FOR ANY I BEAM OR DECK BEAM.

AXIS PARALLEL WITH WEB.

The following rule gives a close approximation for the moment of inertia.

Multiply the increase of area in square inches by the total thickness of web in the enlarged section. This product added to the tabular number in col. 8, will give the moment of inertia for the enlarged section.

Example.—A 10″ I Beam, No. 8, area 9 square inches is increased to $10\frac{1}{2}$ square inches, having a web thickness of .525 inches. $.525 \times 1\frac{1}{2} = .7875$, which added to the amount in col. VIII., $8.09 + .78 = 8.87$, the moment of inertia required.

Radius of gyration of least section $= \sqrt{\dfrac{8.09}{9}} = .95$ inches.

Radius of gyration of enlarged section $= \sqrt{\dfrac{8.87}{10.5}} = .92$ inches.

The radius of gyration alters but very little, and may be accepted as practically unchanged within the limits that any shape can be increased.

CHANNELS.

For channels, in relation to axis parallel to web the moment of inertia increases nearly in a direct ratio to the increase of sectional area, but not precisely so, this ratio being too great for the larger sections and too little for the smaller sizes of channel bars.

The radius of gyration alters but little as the sectional area is changed, and practically may be accepted as unchanged within the range of variation possible for any particular size.

The distance d will not vary sufficiently in any section between the limits of minimum and maximum to make any practical difference in ordinary calculations where it may be used.

ANGLES.

For angles referring to any axis passing through the centre of gravity, the inertia increases nearly in the same ratio as the area increases. Our table gives values of I for the minimum and maximum sections ; any intermediate section can be obtained by proportion unless great accuracy is required. Our tables exhibit the change in values of R between the least and greatest sections, which in the case of small angles remain practically unaltered within the range of possible variation of area.

INERTIA OF COMPOUND SHAPES.

" The moment of inertia of any section about any axis is equal to the I about a parallel axis passing through its centre of gravity + the area of the section multiplied by the square of the distance between the axes."

By use of this rule the moments of inertia or radii of gyration of any single sections being known, corresponding values can readily be obtained for any combination of these sections.

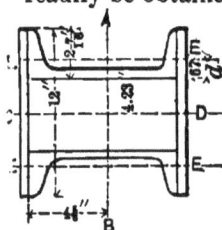

Example No. 1.—A combination of two 9″ 54 lb. Channels, and two 12 × ¼ plates as shown.

AXIS A B OF SECTION.

I for 2 channels, col. VII, page 94,   = 128.680

I for 2 plates $= \dfrac{12 \times .25^3}{12} \times 2 = \quad$ .03125 $\Big\}$

6 (area of plates) × 4⅝ ³   = 128.34375 $\Big\}$ = 128.375

I for combined section   = 257.055

which divided by area (14) gives 18.3611 $= R^2$ or 4.285 radius of combined section.

AXIS C D.

Find distance $d = (.67)$ from col. XV., page 95, then obtaining the distance (4.2325) between axes CD and EF.

I for 2 channels around axis EF from col. VIII.,   = 4.94

Area of channels × square of distance $= 10.8 \times 4.2325^2 = 193.471$

I for 2 plates $= \dfrac{.5 \times 12^3}{12}$ = 72.

I for combined section $= 270.411$

Radius of gyration $= \sqrt{\dfrac{270.411}{14}} = 4.395.$

By similar methods, inertia or radius of gyration for any combination of shapes can readily be obtained.

Example No. 2.—A "built-up beam" composed of :

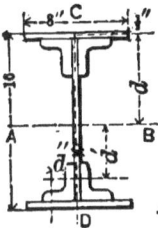

4 angles $3'' \times 3'' \times \frac{1}{4}''$.
2 plates $8'' \times \frac{1}{2}''$.
1 plate $15'' \times \frac{3}{8}''$.

AXIS A B.

I of two $8'' \times \frac{1}{2}$ plates $= \dfrac{8 \times \frac{1}{2}^3}{12} \times 2 =$ .167

$+ \, 8$ (area) $\times \, 7\frac{3}{4}^2$ (sq. of distance d) $= 480.5$

           ——— 480.667

I of one $15'' \times \frac{3}{8}''$ plate $= \dfrac{15^3 \times \frac{3}{8}}{12} =$ 105.469

I of four $3 \times 3 \times \frac{1}{4}$ angles $= 4 \times 1.24$ (see col.
  IV, page 98),   = 4.96
$+ \, 5.77$ (area) $\times \, 6.66^2$ (sq. of distance d') $= 255.045$

           ——— 260.005

Inertia of combined section around $A \, B = 846.141$

Radius of gyration $= \sqrt{\dfrac{846.141}{19.375}} =$ 6.61.

AXIS C. D.

$$I \text{ of two } 8 \times \tfrac{1}{2} \text{ plates} = \frac{8^3 \times \tfrac{1}{2}}{12} \times 2 = \qquad 42.667$$

$$I \text{ of one } 15 \times \tfrac{3}{8} \text{ plate} = \frac{15 \times \tfrac{3}{8}}{12} \qquad = \qquad .066$$

I of four $3 \times 3 \times \tfrac{1}{4}$ angles $= 4 \times 1.24$ (see
 col. IV, page 98) $= \quad 4.96$
$+ 5.75$ (area) $\times 1.0275^2$ (sq. of distance d'') $= \quad 6.071$
              ————— 11.031

Inertia of combined section around $C D = \; 53.764$

$$\text{Radius of gyration} = \sqrt{\frac{53.764}{19.375}} = \; 1.66.$$

RADIUS OF GYRATION OF COMPOUND SHAPES.

In the case of a pair of any shape without a web the value of R can always be readily found without considering the moment of inertia.

The radius of gyration for any section around an axis parallel to another axis passing through its centre of gravity, is found as follows :

Let $r =$ radius of gyration around axis through centre of gravity. $R =$ radius of gyration around another axis parallel to above. $d =$ distance between axes.

$$R = \sqrt{d^2 + r^2}.$$

When r is small, R may be taken as equal to d without material error. Thus in the case of a pair of channels latticed together, or a similar construction.

Example No. 1.—Two 9″ 54 lb. channels placed 4.66″ apart, required the radius of gyration around axis $C D$ for combined section.

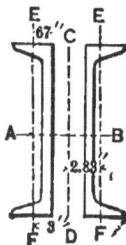

Find r on col. X., page 95, $= .68$ and $r^2 = .4624$.

Find distance from base of channel to neutral axis col. XV., same page, $= .67$, this added to $\tfrac{1}{2}$ distance between the two bars, $2.33'' = 3'' = d$, and $d^2 = 9$.

Radius of gyration of the pair as placed equals,

$$\sqrt{9 + .4624} = 3.076.$$

The value of R for the whole section in relation to the axis A B is the same as for the single channel, to be found in the tables.

Example No 2.—Four $3'' \times 3'' \times \frac{3}{8}''$ angles placed as shown; form a column 10 inches square; required the radius of gyration.

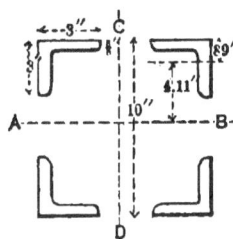

Find r on col. VI, page 98, $= .91$, and $r^2 = .8281$.

Find distance from side of angle to neutral axis, col. VII., same page, $= .89$. Subtract this from $\frac{1}{2}$ the width of column $= 5. - .89 = 4.11 = d$ or distance between two axes. $d^2 = 16.8921.$

Radius of gyration of 4 angles as placed $=$

$$\sqrt{16.8921 + .8281} = 4.21.$$

When the angles are large as compared with the outer dimensions of the combined section, the radius of gyration can be taken without serious error from the table of radii of gyration for square columns, on page 155.

RADIUS OF GYRATION.

The table below exhibits values for the least radius of gyration and the square of the same, in terms of the sides or diameter of the cross section. In most cases the values given are only *approximate*, and those for flanged beams only apply to standard minimum sections. Those marked with an * are theoretically correct.

SHAPE OF SECTION.	SQUARE OF LEAST RADIUS OF GYRATION.	LEAST RADIUS OF GYRATION.
SOLID RECTANGLE.	$\dfrac{(\text{Least side})^2}{12}$ *	$\dfrac{\text{Least side}}{3.46}$ *
THIN HOLLOW SQUARE. a = area inner square. A = area outer square.	$\dfrac{A+a}{12}$ or $\dfrac{d^2+d_1^2}{12}$ *	$\dfrac{d+d_1}{4.89}$
SOLID CIRCLE.	$\dfrac{(\text{Diameter})^2}{16}$ *	$\dfrac{\text{Diameter}}{4}$ *
THIN HOLLOW CIRCLE. a = area inner circle. A = area outer circle.	$\dfrac{A+a}{12.566}$ or $\dfrac{d^2+d_1^2}{16}$ *	$\dfrac{d+d_1}{5.64}$
PHOENIX COLUMN.	$\dfrac{(\text{Diameter})^2}{7.6}$	$\dfrac{\text{Diameter}}{2.75}$

Section		
Angles, Equal Legs.	$\dfrac{(\text{Length of leg})^2}{25}$	$\dfrac{\text{Length of leg}}{5}$
Uneven Angles.	$\dfrac{(l \times l_1)^2}{13(l^2 + l_1^2)}$	$\dfrac{l \times l_1}{2.6\,(l + l_1)}$
Cross, Equal Legs.	$\dfrac{(\text{Greatest Width})^2}{22.5}$	$\dfrac{\text{Greatest Width}}{4.74}$
Tees, Equal Legs.	$\dfrac{(\text{Width of Flange})^2}{22.5}$	$\dfrac{\text{Width of Flange}}{4.74}$
Pencoyd I Beams.	$\dfrac{(\text{Width of Flange})^2}{21}$	$\dfrac{\text{Width of Flange}}{4.58}$
Pencoyd Channels.	$\dfrac{(\text{Width of Flange})^2}{12.5}$	$\dfrac{\text{Width of Flange}}{3.54}$
Pencoyd Deck Beams.	$\dfrac{(\text{Width of Flange})^2}{36.5}$	$\dfrac{\text{Width of Flange}}{6.}$

ROLLED IRON STRUTS.

In the following consideration of rolled struts of various shapes, the least radius of gyration of the cross section taken around an axis through the centre of gravity is assumed as the effective radius of the strut. The resistance of any section per unit of area will in general terms vary directly as the square of the least radius of gyration, and inversely as the square of the length of the strut.* The shape of the section and the distribution of the metal to resist local crippling strains must also be considered. As a rule, that shape will be strongest which presents the least extent of flat unbraced surface. For instance, two ⊢⊣ sections of unequal web widths may have the same web thickness, the same flange area, and the same least radius of gyration, but the wider webbed section will be the weaker per unit of area, on account of the greater extent of unbraced web surface it contains. For the same reason a hollow rectangular section, composed of thin plates will be to some extent weaker than a circular section of the same length having the same area and radius of gyration.

END CONNECTIONS.

As is well known, the method of securing the ends of the struts exercises an important influence on their resistance to bending, as the member is held more or less rigidly in the direct line of thrust.

In the tables, struts are classified in four divisions, viz.: "Fixed Ended," "Flat Ended," "Hinged Ended," and "Round Ended."

In the class of "fixed ends" the struts are supposed to be so rigidly attached at both ends to the contiguous parts of the structure that the attachment would not be severed if the member was subjected to the ultimate load. "Flat ended" struts are supposed to have their ends flat and square with the axis of length but not rigidly attached to the adjoining parts. "Hinged

---

* This applies only to long struts with free ends.

ends " embrace the class which have both ends properly fitted with pins, or ball and socket joints, of substantial dimensions as compared with the section of the strut; the centres of these end joints being practically coincident with an axis passing through the centre of gravity of the section of the strut. " Round ended " struts are those which have only central points of contact, such as balls or pins resting on flat plates, but still the centres of the balls or pins coincident with the proper axis of the strut.

If in hinged-ended struts the balls or pins are of comparatively insignificant diameter, it will be safest in such cases to consider the struts as round ended.

If there should be any serious deviation of the centres of round or hinged ends from the proper axis of the strut, there will be a reduction of resistance that cannot be estimated without knowing the exact conditions. No formula has been written which expresses with accuracy the resistance to compression for various sections and for an extended range of lengths. It is doubtful if any simple formula admitting of ready practical application can be devised; in fact none is required, as the results of experiments can be embodied in tables and diagrams in such a compact form that their application to any length or section can be readily made.

When the pins of hinged-end struts are of substantial diameter, well fitted, and exactly centred, experiment shows that the hinged ended will be equally as strong as flat ended struts.

But a very slight inaccuracy of the centring rapidly reduces the resistance to lateral bending, and as it is almost impossible in practice to uniformly maintain the rigid accuracy required, it is considered best to allow for such inaccuracies to the extent given in the tables, which are the average of many experiments.

TABLES OF STRUTS.

In table No. 1, the first column gives the effective length of the strut divided by the least radius of gyration of its cross section, and the successive columns give the ultimate load per square inch of sectional area for each of the four classes aforesaid. We mean by "ultimate load" that pressure under which the strut fails.

These ultimate loads are the averages of a number of experiments which we have recently made on carefully prepared specimens, and are believed to be trustworthy.

For hinged-ended struts the figures apply to those cases in which the axis of the pin is at right angles to the least radius of gyration, or in which the strut is free to rotate on the pin in its weakest direction. If the pin should be placed in another direction, or if the strut should be secured from failure in its weakest direction, there will be a correction for determining the resistance as hereafter described.

FACTORS OF SAFETY.

It is considered good practice to increase the factors of safety as the length of the strut is increased, owing to the greater inability of the long struts to resist cross strains, etc. For similar reasons we consider it advisable to increase the factor of safety for hinged and round ends in a greater ratio than for fixed or flat ends.

Presuming that one-third of the ultimate load would constitute the greatest safe load for the shortest struts, the following progressive factors of safety are adopted for the increasing lengths.

$$3. + .01\frac{l}{r} \text{ for flat and fixed ends.}$$

$$3 + .015\frac{l}{r} \text{ for hinged and round ends.}$$

$l =$ length of strut.

$r =$ least radius of gyration.

From the above we derive the following table:

FACTORS OF SAFETY.

$\dfrac{l}{r}$	FIXED AND FLAT ENDS.	HINGED AND ROUND ENDS.	$\dfrac{l}{r}$	FIXED AND FLAT ENDS.	HINGED AND ROUND ENDS.	$\dfrac{l}{r}$	FIXED AND FLAT ENDS.	HINGED AND ROUND ENDS.
20	3.2	3.3	150	4.5	5.25	280	5.8	7.2
30	3.3	3.45	160	4.6	5.4	290	5.9	7.35
40	3.4	3.6	170	4.7	5.55	300	6.0	7.5
50	3.5	3.75	180	4.8	5.7	310	6.1	7.65
60	3.6	3.9	190	4.9	5.85	320	6.2	7.8
70	3.7	4.05	200	5.0	6.0	330	6.3	7.95
80	3.8	4.2	210	5.1	6.15	340	6.4	8.1
90	3.9	4.35	220	5.2	6.3	350	6.5	8.25
100	4.0	4.5	230	5.3	6.45	360	6.6	8.4
110	4.1	4.65	240	5.4	6.6	370	6.7	8.55
120	4.2	4.8	250	5.5	6.75	380	6.8	8.7
130	4.3	4.95	260	5.6	6.9	390	6.9	8.85
140	4.4	5.1	270	5.7	7.05	400	7.0	9.0

Table No. 2 represents the greatest safe load per square inch of section for each of the four classes of struts and is derived from the results in Table No. 1 by means of the foregoing factors of safety.

The remarks on page 33 for safe loads on beams, apply also to struts. The loads in Table No. 2 ought to be applied only under the most favorable circumstances, such as an invariable condition of the load, little or no vibration, etc. Under certain conditions, such as for buildings, bridges, etc., the least factor of safety ought to be four (4), which would increase each factor in the above table by unity. The safe load will then be found by dividing the results given in Table No. 1 by the corrected factor of safety.

No. 1.

WROUGHT IRON STRUTS.

ULTIMATE PRESSURE IN LBS. PER SQUARE INCH.

LENGTH / LEAST RADIUS OF GYRATION.	FLAT ENDS.	FIXED ENDS.	HINGED ENDS.	ROUND ENDS.
20	46.000	46,000	46,000	44,000
30	43,000	43,000	43,000	40,250
40	40.000	40,000	40,000	36,500
50	39,000	38.000	38,000	33,500
60	36,000	36,000	36,000	30,500
70	34,000	34,000	33,750	27,750
80	32,000	32,000	31,500	25,000
90	30,900	31,000	29,750	22,750
100	29,800	30,000	28,000	20,500
110	28,050	29,000	26,150	18,500
120	26.300	28,000	24,300	16 500
130	24,900	26,750	22,650	14,650
140	23,500	25,500	21,000	12,800
150	21,750	24,250	18,750	11,150
160	20,000	23,000	16,500	9,500
170	18,400	21,500	14,650	8,500
180	16,800	20,000	12,800	7,500
190	15,650	18,750	11,800	6,750
200	14,500	17,500	10,800	6,000
210	13,600	16.250	9 800	5,500
220	12,700	15,000	8,800	5,000
230	11,950	14,000	8,150	4,650
240	11,200	13,000	7,500	4,300
250	10,500	12,000	7,000	4,050
260	9,800	11,000	6,500	3,800
270	9,150	10,500	6,100	3,500
280	8,500	10,000	5,700	3,200
290	7,850	9,500	5,350	3,000
300	7,200	9 000	5,000	2,800
310	6,600	8,500	4,750	2,650
320	6,000	8.000	4,500	2,500
330	5,550	7,500	4,250	2,300
340	5,100	7,000	4,000	2,100
350	4,700	6,750	3,750	2,000
360	4,300	6,500	3 500	1,900
370	3,900	6,150	3,250	1,800
380	3,500	5,800	3,000	1,700
390	3,250	5,500	2,750	1,600
400	3,000	5,200	2,500	1,500
410	2,750	5,000	2,400	1,400
420	2,500	4,800	2,300	1,300
430	2,350	4,550	2,200	
440	2,200	4,300	2,100	
450	2,100	4,050	2,000	
460	2,000	3,800	1,900	
470	1,950		1,850	
480	1,900		1,800	

No. 2.

GREATEST SAFE LOADS ON STRUTS.

Greatest safe load in lbs. per square inch of cross section for vertical struts. *Both* ends are supposed to be secured as indicated at the head of each column. If both ends are not secured alike, take a mean proportional between the values given for the classes to which each end belongs. If the strut is hinged by any uncertain method so that the centres of pins and axis of strut may not coincide, or the pins may be relatively small and loosely fitted, it is best in such cases to consider the strut as "round ended."

LENGTH / LEAST RADIUS OF GYRATION.	FLAT ENDS.	FIXED ENDS.	HINGED ENDS.	ROUND ENDS.
20	14,380	14,380	13,940	13,330
30	13,030	13,030	12,460	11,670
40	11,760	11,760	11,110	10,140
50	10,860	10,860	10,130	8,930
60	10,000	10,000	9,230	7,820
70	9,190	9,190	8,330	6,850
80	8,420	8,420	7,500	5,950
90	7,920	7,950	6,840	5,230
100	7,450	7,500	6,220	4,560
110	6,840	7,070	5,620	3,980
120	6,260	6,670	5,060	3,440
130	5,790	6,220	4,580	2,960
140	5,340	5,800	4,120	2,510
150	4,830	5,390	3,570	2,120
160	4,350	5,000	3,060	1,760
170	3,920	4,570	2,640	1,530
180	3,500	4,170	2,250	1,310
190	3,190	3,830	2,020	1,150
200	2,900	3,500	1,800	1,000
210	2,670	3,190	1,590	890
220	2,440	2,880	1,400	790
230	2,250	2,640	1,260	720
240	2,070	2,410	1,140	650
250	1,910	2,180	1,040	600
260	1,750	1,960	940	550
270	1,610	1,840	870	500
280	1,460	1,720	790	440
290	1,330	1,610	730	410
300	1,200	1,500	670	370
310	1,080	1,390	620	350
320	970	1,290	580	320
330	880	1,190	540	290
340	800	1,090	490	260
350	720	1,040	450	240
360	650	980	420	230
370	580	920	380	210
380	510	850	340	200
390	470	800	310	80
400	430	740	280	70

ROLLED STRUCTURAL SHAPES AS STRUTS.

The following tables for the working values of various rolled structural shapes as struts are derived directly from Table No. 2. The radii of gyration are taken from Tables of Elements, pages 92–101. In all cases the strut is supposed to stand vertical. In short struts this distinction is immaterial, but when the length becomes considerable, the deflection resulting from its own weight, if horizontal, would seriously affect the stability of the strut.

The tables are calculated for the minimum section of each shape. For sections increased above the minimum the resistance per square inch will diminish. This amount can be accurately determined by finding the correct radius of gyration for the enlarged section as heretofore described. But within the range of variation of section possible for any shape, the tables may be accepted as practically correct. The head notes to the tables indicate the condition assumed for each class of struts. If the pins should be placed otherwise than as described in the tables, the strut may be either weaker or stronger, according to circumstances, which have to be determined for any particular case. This results from the fact that a pin-connected strut if properly designed should be considered hinged ended, only in the direction in which it is free to rotate on the pin.

In the direction of the axis of the pin it can be treated as a "flat ended" strut. An I beam strut of the character described in Tables 3, 4, and 5, braced laterally in the direction of its flanges should be considered also by Tables 6, 7, and 8, as a series of short struts whose lengths are the distances between points of bracing, and liable to fail in the direction of the flanges.

Example.—An 8″ 65 lb. I beam, 18 feet long is used as a strut having pins at both ends at right angles to web. It would then be flat ended in the direction of the flanges, and by Table No. 7 the greatest safe load = 1,990 lbs. per square inch of section. If braced in the direction of the flanges at two points 6 feet apart it should then be considered as a series of flat ended struts 6 feet long, whose safe load by Table No. 7, would be 8,320 lbs. per square inch.

In the direction of its web it remains a hinged-ended strut 18 feet long, and safe load by Table No. 4 = 8,690 lbs. per square inch.

CHANNEL STRUTS.

The foregoing remarks apply also to channels, which are seldom used individually as struts, but frequently in pairs. When so used, if the methods of connection are not of such a nature as to insure the unity of action of the pair, they should be treated as an assemblage of separate struts. But if connected by a proper system of triangular latticing, the pair can be considered as a unit, and each channel treated as a series of short struts whose length is the distance between centres of latticing.

Example.—A pair of 9″ 54 lb. channels, separated, etc., as described on page 110, are connected by triangular latticing, forming a hinged-ended strut 10 feet between pin centres. What is the greatest safe load, and how far can latticing be spaced ?

As described on page 95, radius of gyration around axis across the web of channel, or in the direction of the pin = 3.45 inches. Radius of gyration in opposite direction = 3.07 inches. Least radius of gyration for a single channel = .68 inch.

$\frac{l}{r}$ for hinged-ended direction = 35, and by Table No. 2 Safe Load = 11,800 lbs. $\frac{l}{r}$ for flat-ended direction = 39, and by same table greatest safe load = 11,900 lbs.

For each single channel the greatest length between latticing = radius of gyration $\times$ 39 = 26¼ inches.

It is customary and is also good practice to reduce the distance between lattice centres below what the above calculation would require.

Tables Nos. 12–14, give the greatest safe loads per square inch of sectional areas, for struts composed of a pair of channels properly connected together, so as to insure unity of action. The figures are derived from Table No. 2.

The distances D or d, for channels placed flanges inward or flanges outward respectively, make the radii of gyration equal for either direction of axis.

These distances should not be diminished, and may be advantageously increased, especially for hinged-ended struts, if the pin is placed parallel to the webs of the channels. These tables are calculated for the standard minimum section of each channel. The distance d may be slightly diminished for sections heavier than the minimum, but the diminution can be so little that it is practically unnecessary to notice it. Under each length of struts in the table l represents the greatest distance apart in feet that centres of lateral bracing can be spaced, without allowing weakness in the individual channels. The distance l is obtained as shown in last example, that is, by making $\dfrac{l}{r} = \dfrac{L}{R}$.

$l =$ length between bracing.
$L =$ total length of strut.
$r =$ least radius of gyration for a single channel.
$R =$ least radius of gyration for the whole section.

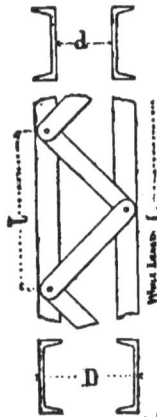

STEEL STRUTS.

A table for the ultimate resistance of flat-ended struts of two grades of steel will be found on page 31. These grades probably embrace the extremes of the material, that is, the hardest and softest steels that are likely to be used in struts.

Experiments on this material are not sufficiently complete to warrant a full statement of resistances of the various grades, and for the various conditions of the strut, such as the methods of connecting the ends, etc.

It is probable, however, that the relations existing between the four classes of wrought-iron struts, as given in the following tables, will also prevail in the same ratios for steel. The safe loads for steel struts of any section or length, can therefore be obtained by increasing the figures in the following tables, for any ratio of $\dfrac{l}{r}$, in the proportions given on page 31, as existing between flat-ended struts of iron and steel.

When a grade of steel is used, intermediate in hardness between the mild and hard heretofore described, it is probable that the strut resistance for such material may be safely approximated by simple proportion.

For instance, the steels referred to had carbon ratios of .12 and .36 per cent. respectively. A mean proportion of these would be .24 per cent.

It is probable that steel of latter grade would possess intermediate compressive resistance between the two grades described from our experiments.

No. 3.

PENCOYD I BEAMS AS STRUTS.

GREATEST SAFE LOAD IN LBS. PER SQUARE INCH OF SECTION.

When the struts are secure from failure in the direction of the flanges, and can bend only in the direction of the web C D. Using factors of safety given in previous tables.

SIZE OF BEAM.	CONDITION OF ENDS.	LENGTH IN FEET.								
		8	10	12	14	16	18	20	22	24
15″ Heavy.... r = 6.86	Fixed Ends.. ..	.....	14240	13700	13160	12650	12140	11670	11310	10950
	Flat Ends......	.....	14240	13700	13160	12650	12140	11670	11310	10950
	Hinged Ends...	.....	13790	13200	12610	12050	11510	11010	10620	10230
	Round Ends....	.....	13160	12500	11840	11210	10600	10020	9530	9050
15″ Light.... r = 6.98	Fixed Ends....	.....	14380	13840	13300	12780	12270	11760	11400	11040
	Flat Ends......	.....	14380	1384')	13300	12780	12270	11760	11400	11040
	Hinged Ends...	.....	13940	13350	12760	12190	11650	11110	10720	10330
	Round Ends....	.....	13330	12670	12000	11360	10750	10140	9660	9170
12″ Heavy.... r = 4.69	Fixed Ends....	14380	13570	12900	12270	11670	11220	10770	10340	9920
	Flat Ends......	14380	13570	12900	12270	11670	11220	10770	10340	9920
	Hinged Ends...	13940	13050	12320	11650	11010	10520	10040	9590	9140
	Round Ends....	13330	12330	11520	10750	10020	9410	8820	8260	7720
12″ Light.... r = 4.78	Fixed Ends....	14380	13700	13030	12400	11760	11310	10860	10430	10000
	Flat Ends......	14380	13700	13030	12400	11760	11310	10860	10430	10000
	Hinged Ends...	13940	13200	12460	11780	11110	10620	10130	9680	9230
	Round Ends....	13330	12500	11670	10900	10140	9530	8930	8370	7820
10½″ Heavy.... r = 4.34	Fixed Ends. .	13970	13160	12400	11760	11220	10690	10170	9760	9270
	Flat Ends......	1390	13160	12400	11760	11220	10690	10170	9760	9270
	Hinged Ends...	13500	12610	11780	11110	10520	9950	9410	8960	8420
	Round Ends....	12830	11840	10900	10140	9410	8710	8040	7530	6950
10½″ Light.... r = 4.24	Fixed Ends....	13970	13160	12400	11760	11220	10690	10170	9760	9270
	Flat Ends......	13970	13160	12400	11760	11220	10690	10170	9760	9270
	Hinged Ends...	13500	12610	11780	11110	10520	9950	9410	8960	8420
	Round Ends....	12830	11840	10900	10140	9410	8710	8040	7530	6950
10″ Heavy.... r = 3.94	Fixed Ends....	13840	13030	12140	11490	10950	10430	9920	9430	8960
	Flat Ends......	13840	13030	12140	11490	10950	10430	9920	9430	8960
	Hinged Ends...	13350	12460	11510	10820	10230	9680	9140	8600	8090
	Round Ends....	12670	11670	10600	9780	9050	8370	7720	7140	6580

No. 3.

PENCOYD I BEAMS AS STRUTS.

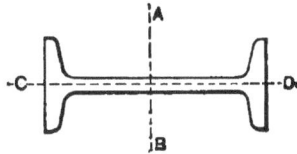

In the marginal columns r indicates the radius of gyration taken around axis $A\,B$. When strut is hinged the pins are supposed to lie in the direction $A\,B$. Under the conditions stated the strut may be considered flat ended in direction $A\,B$.

LENGTH IN FEET.									CONDITION OF ENDS.	SIZE OF BEAM.
26	28	30	32	34	36	38	40	42		
10600	10260	9920	9510	9190	8880	8580	8330	8140	Fixed Ends....	**15″**
10600	10260	9920	9510	9190	8880	8580	8320	8120	Flat Ends......	
9860	9500	9140	8690	8330	8000	7670	7370	7100	Hinged Ends...	Heavy.
8600	8150	7720	7240	6850	6490	6130	5810	5420	Round Ends....	$r = 5.86$
10690	10340	10000	9680	9350	9040	8730	8420	8230	Fixed Ends....	**15″**
10690	10340	10000	9680	9350	9040	8730	8420	8220	Flat Ends.	
9950	9590	9230	8870	8510	8160	7830	7500	7240	Hinged Ends...	Light.
8710	8260	7820	7430	7040	6670	6310	5950	5660	Round Ends....	$r = 5.98$
9430	9040	8650	8330	8090	7860	7630	7410	7190	Fixed Ends....	**12″**
9430	9040	8650	8320	8070	7830	7590	7330	7020	Flat Ends	
8600	8160	7750	7370	7040	6720	6410	6100	5800	Hinged Ends...	Heavy.
7140	6670	6220	5810	5350	5100	4760	4440	4150	Round Ends....	$r = 4.69$
9590	9190	8800	8420	8180	7950	7720	7500	7280	Fixed Ends....	**12″**
9590	9190	8800	8420	8170	7920	7680	7450	7140	Flat Ends......	
8780	8330	7910	7500	7170	6840	6530	6220	5920	Hinged Ends...	Light.
7330	6850	6400	5950	5490	5230	4890	4560	4270	Round Ends....	$r = 4.78$
8800	8420	8140	7860	7590	7320	7070	6870	6620	Fixed Ends....	**10½″**
8800	8420	8120	7830	7540	7210	6840	6550	6210	Flat Ends......	
7910	7500	7100	6720	6340	5980	5620	5340	5010	Hinged Ends...	Heavy.
6400	5950	5420	5100	4690	4320	3980	3710	3390	Round Ends....	$r = 4.24$
8800	8420	8090	7810	7540	7320	7070	6870	6620	Fixed Ends....	**10½″**
8800	8420	8070	7780	7500	7210	6840	6550	6210	Flat Ends......	
7910	7500	7040	6650	6280	5980	5620	5340	5010	Hinged Ends...	Light.
6400	5950	5350	5030	4630	4320	3980	3710	3390	Round Ends....	$r = 4.26$
8500	8180	7900	7630	7320	7070	6830	6580	6310	Fixed Ends....	**10″**
8500	8170	7870	7590	7210	6840	6490	6160	5880	Flat Ends......	
7580	7170	6780	6410	5980	5620	5280	4960	4670	Hinged Ends...	Heavy.
6040	5490	5160	4760	4320	3980	3650	3340	3050	Round Ends....	$r = 3.94$

No. 4.

PENCOYD I BEAMS AS STRUTS.

GREATEST SAFE LOAD IN LBS. PER SQUARE INCH OF SECTION.

When the struts are secure from failure in the direction of the flanges and can bend only in the direction of the web CD. Using factors of safety given in previous tables.

SIZE OF BEAM.	CONDITION OF ENDS.	6	8	10	12	14	16	18	20	22
10″ Light r = 4·05	Fixed Ends....		13840	13030	12270	11670	11130	10600	10090	9590
	Flat Ends....		13840	13030	12270	11670	11130	10600	10090	9590
	Hinged Ends...		13350	12460	11650	11010	10420	9860	9320	8780
	Round Ends....		12670	11670	10750	10020	9290	8600	7930	7330
9″ Heavy r = 3·62	Fixed Ends....	14380	13430	12650	11760	11130	10600	10000	9510	8960
	Flat Ends....	14380	13430	12650	11700	11130	10600	10000	9510	8960
	Hinged Ends...	13940	12900	12050	11110	10420	9860	9230	8660	8080
	Round Ends....	13330	12170	11210	10140	9290	8600	7820	7240	6580
9″ Light r = 3·66	Fixed Ends....	14380	13570	12650	11890	11220	10690	10090	9590	9040
	Flat Ends....	14380	13570	12650	11890	11220	10690	10090	9590	9040
	Hinged Ends...	13940	13050	12050	11240	10520	9950	9320	8780	8160
	Round Ends....	13330	12330	11210	10290	9410	8710	7930	7330	6670
8″ Heavy r = 3·21	Fixed Ends...	14110	13030	12140	11310	10690	10000	9430	8800	8330
	Flat Ends....	14110	13030	12140	11310	10690	10000	9430	8800	8320
	Hinged Ends...	13640	12460	11510	10620	9950	9230	8600	7910	7370
	Round Ends....	13000	11670	10600	9590	8710	7820	7140	6400	5810
8″ Light r = 3·26	Fixed Ends....	14110	13160	12140	11400	10690	10090	9510	8880	8370
	Flat Ends....	14110	13160	12140	11400	10690	10090	9510	8880	8370
	Hinged Ends...	13640	12610	11510	10720	9950	9320	8690	8000	7430
	Round Ends....	13000	11840	10000	9660	8710	7930	7240	6490	5880
7″ Heavy r = 2·76	Fixed Ends....	13570	12400	11400	10690	9920	9190	8500	8090	7380
	Flat Ends....	13570	12400	11400	10690	9920	9190	8500	8070	7640
	Hinged Ends...	13050	11780	10720	9950	9140	8330	7580	7040	6470
	Round Ends....	12330	10900	9660	8710	7720	6850	6040	5350	4830
7″ Light r = 2·89	Fixed Ends....	13700	12650	11580	10860	10170	9510	8800	8280	7900
	Flat Ends....	13700	12650	11580	10860	10170	9510	8800	8270	7870
	Hinged Ends...	13200	12050	10910	10130	9410	8690	7910	7300	6780
	Round Ends....	12500	11210	9900	8930	8040	7240	6400	5730	5160

No. 4.

PENCOYD I BEAMS AS STRUTS.

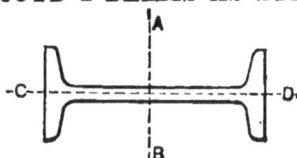

In the marginal columns *r* indicates the radius of gyration taken around axis *A B*. When strut is hinged the pins are supposed to lie in the direction *A B*. Under the conditions stated the strut may be considered flat ended in direction *A B*.

\-\-				LENGTH IN FEET.					CONDITION OF ENDS.	SIZE OF BEAM.
24	26	28	30	32	34	36	38	40		
9110	8650	8280	8000	7720	7450	7190	6990	6750	Fixed Ends....	**10″**
9110	8650	8270	7970	7680	7390	7020	6720	6370	Flat Ends......	
8250	7750	7300	6910	6530	6160	5800	5500	5170	Hinged Ends...	Light.
6760	6220	5730	5200	4890	4500	4150	3870	3540	Round Ends ...	r = 4·05
8420	8140	7810	7540	7240	6950	6710	6400	6090	Fixed Ends....	**9″**
8420	8120	7780	7500	7080	6660	6310	5970	5650	Flat Ends......	
7500	7100	6650	6280	5860	5450	5110	4770	4440	Hinged Ends..	Heavy.
5950	5420	5030	4630	4210	3810	3490	3150	2820	Round Ends....	r = 3·62
8590	8180	7900	7590	7320	7030	6790	6490	6170	Fixed Ends....	**9″**
8580	8170	7870	7540	7210	6780	6430	6070	5740	Flat Ends......	
7670	7110	6780	6340	5980	5560	5220	4860	4530	Hinged Ends...	Light.
6130	5490	5160	4690	4320	3920	3600	3240	2910	Round Ends....	r = 3·63
7950	7630	7280	6990	6670	6350	6010	5710	5430	Fixed Ends....	**8″**
7930	7500	7140	6720	6360	5930	5560	5230	4880	Flat Ends......	
6840	6410	5920	5500	5060	4720	4350	4010	3620	Hinged Ends...	Heavy.
5230	4760	4270	3870	3440	3100	2730	2430	2150	Round Ends....	r = 3·21
8000	7680	7320	7030	6750	6400	6090	5800	5510	Fixed Ends...	**8″**
7970	7640	7210	6780	6370	5970	5650	5340	4980	Flat Ends......	
6910	6470	5990	5560	5170	4770	4440	4120	3730	Hinged Ends...	Light.
5200	4830	4330	3920	3540	3150	2820	2510	2230	Round Ends....	r = 3·26
7280	6950	6580	6170	5800	5470	5110	4740	4370	Fixed Ends...	**7″**
7110	6660	6160	5740	5340	4930	4490	4090	3710	Flat Ends.	
5920	5450	4960	4530	4120	3680	3210	2800	2440	Hinged Ends...	Heavy.
4270	3810	3340	2910	2510	2190	1860	1620	1420	Round Ends....	r = 2·75
7500	7150	6830	6440	6090	5750	5430	5070	4740	Fixed Ends....	**7″**
7450	6960	6490	6020	5650	5280	4880	4440	4090	Flat Ends......	
6220	5740	5280	4820	4440	4060	3620	3160	2800	Hinged Ends...	Light.
4560	4090	3650	3200	2820	2470	2150	1830	1620	Round Ends...	r = 2·69

No. 5.

PENCOYD I BEAMS AS STRUTS.

GREATEST SAFE LOAD IN LBS. PER SQUARE INCH OF SECTION.

When the struts are secure from failure in the direction of the flanges, and can bend only in the direction of the web *C. D.* Using factors of safety given in previous tables.

SIZE OF BEAM.	CONDITION OF ENDS.	LENGTH IN FEET.								
		2	4	6	8	10	12	14	16	18
6'' Heavy r = 2·31	Fixed Ends....		14240	12900	11580	10690	9840	8960	8280	7810
	Flat Ends.....		14240	12900	11580	10690	9840	8960	8270	7780
	Hinged Ends...		13790	12320	10910	9950	9050	8080	7300	6650
	Round Ends....		13160	11520	9900	8710	7630	6580	5730	5030
6'' Light r = 2·43	Fixed Ends ...		14380	13030	11760	10950	10090	9270	8500	8000
	Flat Ends....		14380	13030	11760	10950	10090	9270	8500	7970
	Hinged Ends...		13940	12460	11110	10230	9320	8420	7580	6910
	Round Ends....		13330	11670	10140	9050	7930	6950	6040	5200
5'' Heavy r = 1·99	Fixed Ends....		13840	12270	11040	10000	9040	8230	7680	7110
	Flat Ends.....		13840	12270	11040	10000	9040	8220	7640	6900
	Hinged Ends...		13350	11650	10330	9230	8160	7240	6470	5680
	Round Ends....		12670	10750	9170	7820	6670	5660	4830	4030
5'' Light r = 2·06	Fixed Ends....		14110	12650	11400	10340	9430	8580	8000	7500
	Flat Ends.....		14110	12650	11400	10340	9430	8580	7970	7450
	Hinged Ends...		13640	12050	10720	9590	8600	7670	6910	6220
	Round Ends ...		13000	11210	9660	8260	7140	6130	5200	4560
4'' Heavy r = 1·63	Fixed Ends....		13160	11400	10090	8880	8040	7370	6750	6130
	Flat Ends.....		13160	11400	10090	8880	8020	7270	6370	5700
	Hinged Ends..		12610	10720	9320	8000	6970	6040	5170	4480
	Round Ends....		11840	9660	7930	6490	5270	4380	3540	2870
4'' Light r = 1·65	Fixed Ends...		13160	11400	10170	8960	8090	7410	6830	6170
	Flat Ends.....		13160	11400	10170	8960	8070	7330	6490	5740
	Hinged Ends.		12610	10720	9410	8080	7040	6100	5280	4530
	Round Ends....		11840	9660	8040	6580	5350	4440	3650	2910
3'' Heavy r = 1·21	Fixed Ends....	14380	11760	10000	8500	7540	6710	5840	5080	4210
	Flat Ends......	14380	11760	10000	8500	7500	6310	5380	4390	3540
	Hinged Ends...	13940	11110	9230	7580	6280	5110	4160	3110	2280
	Round Ends....	13330	10140	7820	6040	4630	3490	2550	1790	1350
3'' Light r = 1·26	Fixed Ends....	14520	12010	10170	8650	7680	6870	6050	5230	4450
	Flat Ends.....	14520	12010	10170	8650	7640	6550	5610	4630	3790
	Hinged Ends...	14090	11380	9410	7750	6470	5340	4390	3360	2520
	Round Ends....	13500	10440	8040	6220	4830	3710	2780	1970	1460

No. 5.

PENCOYD I BEAMS AS STRUTS.

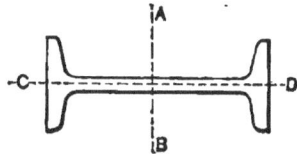

In the marginal columns r indicates the radius of gyration taken around axis $A. B.$ When strut is hinged the pins are supposed to lie in the direction $A. B.$ Under the conditions stated the strut may be considered flat ended in direction $A. B.$

LENGTH IN FEET.									CONDITION OF ENDS.	SIZE OF BEAM.
20	22	24	26	28	30	32	34	36		
7320	6910	6440	6010	5590	5150	4740	4290	3930	Fixed Ends....	6''
7210	6600	6020	5560	5080	4540	4090	3620	3280	Flat Ends......	
5980	5390	4820	4350	3840	3260	2800	2360	2080	Hinged Ends...	Heavy.
4320	3760	3200	2730	2310	1900	1620	1370	1190	Round Ends ...	$r = 2·31$
7540	7110	6710	6310	5880	5470	5070	4650	4250	Fixed Ends....	6''
7500	6900	6310	5880	5430	4930	4440	4000	3580	Flat Ends......	
6280	5680	5110	4670	4210	3680	3160	2720	2320	Hinged Ends...	Light.
4630	4030	3490	3050	2600	2190	1830	1570	1350	Round Ends....	$r = 2·43$
6620	6090	5590	5110	4610	4130	3730	3340	2970	Fixed Ends....	5''
6210	5650	5080	4490	3960	3460	3100	2780	2500	Flat Ends......	
5010	4440	3840	3210	2680	2220	1950	1690	1450	Hinged Ends...	Heavy.
3390	2820	2310	1860	1550	1290	1100	940	820	Round Ends....	$r = 1·99$
7030	6580	6090	5630	5150	4690	4250	3860	3500	Fixed Ends....	5''
6780	6160	5650	5130	4540	4040	3580	3220	2900	Flat Ends......	
5560	4960	4440	3900	3260	2760	2320	2040	1800	Hinged Ends...	Light.
3920	3340	2820	2350	1900	1590	1350	1160	1000	Round Ends....	$r = 2·06$
5510	4910	4290	3790	3310	2850	2500	2180	1900	Fixed Ends....	4''
4980	4260	3620	3160	2760	2420	2140	1910	1680	Flat Ends......	
3730	2970	2360	1990	1670	1380	1180	1040	900	Hinged Ends...	Heavy.
2230	1710	1370	1130	930	780	670	600	520	Round Ends....	$r = 1·63$
5590	5000	4410	3860	3400	2940	2570	2240	1930	Fixed Ends....	4''
5080	4350	3750	3220	2830	2480	2190	1950	1720	Flat Ends......	
3640	3060	2480	2040	1730	1430	1220	1070	920	Hinged Ends...	Light.
2310	1760	1440	1160	960	810	690	610	540	Round Ends....	$r = 1·65$
3560	2940	2450	2000	1740	1520	1320	1120	990	Fixed Ends....	3''
2950	2480	2100	1780	1490	1220	1000	820	670	Flat Ends......	
1840	1430	1160	960	800	680	590	500	420	Hinged Ends...	Heavy.
1030	810	660	560	450	370	320	260	230	Round Ends....	$r = 1·21$
3760	3150	2610	2180	1850	1630	1420	1220	1060	Fixed Ends....	3''
3130	2640	2230	1910	1620	1350	1110	900	750	Flat Ends......	
1970	1570	1240	1040	870	740	630	550	460	Hinged Ends...	Light.
1120	880	710	600	500	410	350	290	240	Round Ends....	$r = 1·26$

9

No 6.

PENCOYD I BEAMS AS STRUTS.

GREATEST SAFE LOAD IN LBS. PER SQUARE INCH OF SECTION.

When the struts are free to bend at right angles to the web ; or in the weakest direction *C. D.* Using factors of safety given in previous tables.

SIZE OF BEAM.	CONDITION OF ENDS.	LENGTH IN FEET.								
		2	4	6	8	10	12	14	16	18
15" Heavy.... r = 1·20	Fixed Ends....	14380	11760	10000	8420	7500	6670	5800	5000	4170
	Flat Ends......	14380	11760	10000	8420	7450	6260	5340	4350	3500
	Hinged Ends...	13940	11110	9230	7500	6220	5060	4120	3060	2250
	Round Ends....	13330	10140	7820	5950	4560	3440	2510	1760	1310
15" Light..... r = 1·08	Fixed Ends....	14110	11400	9430	8000	7030	6090	5150	4250	3500
	Flat Ends......	14110	11400	9430	7970	6780	5650	4540	3580	2900
	Hinged Ends...	13640	10720	8600	6910	5560	4440	3260	2320	1800
	Round Ends....	13000	9660	7140	5200	3920	2820	1900	1350	1000
12" Heavy.... r = 1·17	Fixed Ends....	14240	11670	9840	8330	7370	6530	5630	4820	4000
	Flat Ends......	14240	11670	9840	8320	7270	6110	5130	4170	3340
	Hinged Ends...	13790	11010	9050	7370	6040	4910	3900	2890	2130
	Round Ends....	13160	10020	7630	5810	4380	3290	2350	1660	1230
12" Light .... r = 1·01	Fixed Ends....	13840	11040	9110	7720	6710	5670	4740	3830	3060
	Flat Ends......	13840	11040	9110	7680	6310	5180	4090	3190	2570
	Hinged Ends...	13350	10330	8250	6530	5110	3950	2800	2020	1510
	Round Ends....	12670	9170	6760	4890	3490	2390	1620	1150	850
10½" Heavy,... r = 1·19	Fixed Ends....	14380	11760	10000	8370	7450	6620	5770	4950	4130
	Flat Ends......	14330	11760	10000	8370	7390	6210	5280	4300	3460
	Hinged Ends...	13910	11110	9230	7430	6160	5010	4060	3010	2220
	Round Ends....	13330	10140	7820	5880	4500	3390	2470	1730	1290
10½" Light..... r = ·97	Fixed Ends....	13840	10950	8960	7590	6580	5510	4530	3630	2880
	Flat Ends......	13840	10950	8960	7540	6160	4980	3870	3010	2440
	Hinged Ends...	13350	10230	8080	6340	4960	3730	2600	1880	1400
	Round Ends....	12670	9050	6580	4690	3340	2230	1500	1060	790
10" Heavy.... r = ·98	Fixed Ends ...	13840	10950	8960	7590	6580	5510	4530	3630	2880
	Flat Ends......	13840	10950	8960	7540	6160	4980	3870	3010	2440
	Hinged Ends...	13350	10230	8080	6340	4960	3730	2600	1880	1400
	Round Ends....	12670	9050	6580	4690	3340	2230	1500	1060	790

No. 6.
PENCOYD I BEAMS AS STRUTS.

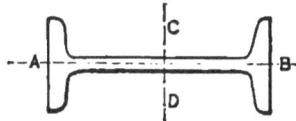

In the marginal columns r indicates the radius of gyration taken around axis A. B. When the strut is hinged the pins are supposed to lie in the direction A. B. If the pins lie in the direction C. D. consider the strut flat ended by this table.

LENGTH IN FEET.									CONDITION OF ENDS.	SIZE OF BEAM.
20	22	24	26	28	30	32	34	36		
3500	2880	2410	1960	1720	1500	1290	1090	980	Fixed Ends....	15″
2900	2440	2070	1750	1460	1200	970	800	650	Flat Ends......	
1800	1400	1140	940	790	670	580	490	420	Hinged Ends...	Heavy.
1000	790	650	550	440	370	320	260	230	Round Ends....	$r = 1 \cdot 20$
2830	2310	1870	1620	1380	1160	1000	860	740	Fixed Ends....	15″
2400	2000	1650	1340	1060	850	670	520	430	Flat Ends......	
1370	1100	890	730	610	520	430	340	280	Hinged Ends...	Light.
770	630	510	410	340	280	230	200	170	Round Ends....	$r = 1 \cdot 09$
3340	2730	2270	1870	1640	1410	1210	1040	920	Fixed Ends....	12″
2780	2320	1970	1650	1360	1100	890	720	580	Flat Ends.....	
1690	1310	1080	890	740	630	540	450	380	Hinged Ends...	Heavy.
940	740	620	510	410	350	290	240	210	Round Ends....	$r = 1 \cdot 17$
2450	1940	1660	1400	1160	1000	850	720	.....	Fixed Ends....	12″
2100	1730	1390	1090	850	670	510	410	.....	Flat Ends......	
1160	930	760	620	520	430	340	270	.....	Hinged Ends...	Light.
660	540	420	350	280	230	200	160	.....	Round Ends....	$r = 1 \cdot 01$
3430	2830	2360	1930	1690	1470	1260	1070	960	Fixed Ends....	10½″
2850	2400	2030	1720	1430	1170	940	770	620	Flat Ends......	
1750	1370	1120	920	770	660	560	470	400	Hinged Ends...	Heavy.
970	770	640	540	430	360	310	250	220	Round Ends....	$r = 1 \cdot 19$
2290	1850	1560	1310	1070	930	780	670	.....	Fixed Ends....	10½″
1990	1620	1270	990	770	600	460	380	.....	Flat Ends......	
1090	870	700	580	470	390	300	240	.....	Hinged Ends...	Light.
620	500	390	320	250	210	170	150	.....	Round Ends....	$r = \cdot 97$
2290	1850	1560	1310	1070	930	780	670	.....	Fixed Ends....	10″
1990	1620	1270	990	770	600	460	380	.....	Flat Ends......	
1000	870	700	580	470	390	300	240	.....	Hinged Ends...	Heavy.
620	500	390	320	250	210	170	150	.....	Round Ends....	$r = \cdot 98$

No. 7.

PENCOYD I BEAMS AS STRUTS.

GREATEST SAFE LOAD IN LBS. PER SQUARE INCH OF SECTION.

The strut is supposed to be free to bend in the weakest direction $C. D.$
The radius of gyration is taken around $A. B.$

SIZE OF BEAM.	CONDITION OF ENDS.	LENGTH IN FEET.								
		2	4	6	8	10	12	14	16	18
10″ Light .... r = ·95	Fixed Ends....	13700	10770	8730	7450	6400	5310	4290	3430	2710
	Flat Ends......	13700	10770	8730	7390	5970	4730	3620	2850	2300
	Hinged Ends...	13200	10040	7830	6160	4770	3460	2360	1750	1300
	Round Ends....	12500	8820	6310	4500	3150	2040	1370	970	740
9″ Heavy.... r = ·94	Fixed Ends....	13700	10860	8800	7500	6440	5390	4370	3500	2760
	Flat Ends......	13700	10860	8800	7450	6020	4830	3710	2900	2340
	Hinged Ends...	13200	10130	7910	6220	4820	3570	2440	1800	1330
	Round Ends....	12500	8930	6400	4560	3200	2120	1420	1000	750
9″ Light.... r = ·89	Fixed Ends....	13430	10530	8370	7150	6010	4910	3860	3000	2340
	Flat Ends......	13430	10520	8370	6960	5560	4260	3220	2530	2020
	Hinged Ends...	12900	9770	7430	5740	4350	2970	2040	1470	1110
	Round Ends....	12170	8490	5880	4090	2730	1710	1160	830	630
8″ Heavy.... r = ·94	Fixed Ends....	13570	10770	8650	7410	6310	5270	4210	3370	2640
	Flat Ends......	13570	10770	8650	7330	5880	4680	3540	2800	2250
	Hinged Ends...	13050	10040	7750	6100	4670	3410	2290	1710	1260
	Round Ends....	12330	8820	6220	4440	3050	2010	1330	950	720
8″ Light.... r = ·89	Fixed Ends....	13430	10430	8330	7110	5960	4820	3790	2940	2290
	Flat Ends......	13430	10430	8320	6900	5520	4170	3160	2480	1990
	Hinged Ends...	12900	9680	7370	5680	4300	2890	1990	1430	1090
	Round Ends....	12170	8370	5810	4030	2690	1660	1130	810	620
7″ Heavy.... r = ·79	Fixed Ends....	13030	9990	7900	6620	5310	4100	3090	2340	1800
	Flat Ends......	13030	9920	7870	6210	4730	3430	2600	2020	1560
	Hinged Ends...	12460	9140	6780	5010	3460	2200	1530	1110	840
	Round Ends....	11670	7720	5160	3390	2040	1270	860	630	480
7″ Light.... r = ·82	Fixed Ends....	13160	10090	8040	6790	5550	4330	3340	2540	1920
	Flat Ends......	13160	10090	8020	6430	5030	3680	2780	2170	1700
	Hinged Ends...	12610	9320	6970	5220	3790	2400	1690	1210	910
	Round Ends....	11840	7930	5270	3600	2270	1390	940	690	530

No. 7.

PENCOYD I BEAMS AS STRUTS.

A. B. indicates the direction of pins for hinged struts in this table. If the pins are placed in the direction *C. D.* consider the strut as flat ended. *r* in marginal columns indica.es radius of gyration around *A. B.*

LENGTH IN FEET.									CONDITION OF ENDS.	SIZE OF BEAM.
20	22	24	26	28	30	32	34	36		
2110	1740	1460	1210	1020	850	720	.....	.....	Fixed Ends....	**10″**
1860	1490	1160	890	690	510	410	.....	.....	Flat Ends......	
1010	800	650	540	440	340	270	.....	.....	Hinged nds...	Light.
580	450	360	290	230	200	160	.....	.....	Round Ends....	r ⁼ ·95
2180	1780	1500	1240	1040	880	740	.....	.....	Fixed Ends....	**9″**
1910	1530	1200	920	720	540	430	.....	.....	Flat Ends......	
1040	830	670	560	450	360	280	.....	.....	Hinged Ends...	Heavy.
600	470	370	300	240	200	170	.....	.....	Round Ends....	r ⁼ ·96
1840	1530	1250	1030	860	720	.....	.....	.....	Fixed Ends....	**9″**
1610	1230	930	710	520	410	. ...	.....	.....	Flat Ends......	
870	680	560	440	340	270	.....	.....	.....	Hinged Ends...	Light.
500	380	300	230	200	160	.....	.....	.....	Round Ends....	r ⁼ ·89
2070	1700	1430	1170	990	830	700	.....	.....	Fixed Ends....	**8″**
1830	1440	1120	860	670	490	400	.....	.....	Flat Ends......	
990	780	640	530	420	330	250	.....	.....	Hinged Ends...	Heavy.
570	430	350	280	230	190	160	.....	.....	Round Ends....	r ⁼ ·94
1800	1500	1220	1010	840	700	.....	.....	.....	Fixed Ends....	**8″**
1560	1200	900	680	500	400	.....	.....	.....	Flat Ends......	
840	670	550	430	330	250	.....	.....	.....	Hinged Ends...	Light.
480	370	290	230	190	160	.....	.....	.....	Round Ends....	r ⁼ ·88
1450	1150	950	770	.....	.....	.....	.....	.....	Fixed Ends....	**7″**
1150	840	610	450	.....	.....	.....	.....	.....	Flat Ends......	
650	520	400	290	.....	.....	.....	.....	.....	Hinged Ends...	Heavy.
360	270	220	170	.....	.....	.....	.....	.....	Round Ends....	r ⁼ ·79
1570	1270	1030	850	700	.....	.....	.....	.....	Fixed Ends....	**7″**
1290	950	710	510	400	.....	.....	.....	.....	Flat Ends......	
710	570	440	340	250	.....	.....	.....	.....	Hinged Ends...	Light.
390	310	230	200	160	.....	.....	.....	.....	Round Ends....	r ⁼ ·82

No. 8.

PENCOYD I BEAMS AS STRUTS.

GREATEST SAFE LOAD IN LBS. PER SQUARE INCH OF SECTION.

(See remarks at head of Tables No. 6 and 7.)

Size of Beam	Condition of Ends.	Length in Feet.								
		2	4	6	8	10	12	14	16	18
6″ Heavy.... r = ·55	Fixed Ends....	12140	8880	7030	5470	4000	2830	2000	1550	1170
	Flat Ends......	12140	8880	6780	4930	3340	2400	1780	1260	860
	Hinged Ends...	11510	8000	5560	3680	2130	1370	960	700	530
	Round Ends....	10600	6490	3920	2190	12.0	770	560	390	280
6″ Light .... r = ·66	Fixed Ends....	12270	8960	7110	5590	4100	2940	2090	1590	1220
	Flat Ends......	12270	8960	6900	5080	3430	2480	1840	1310	900
	Hinged Ends...	11650	8080	5680	3840	2200	1430	1000	720	550
	Round Ends....	10750	6580	4030	2310	1270	810	580	400	290
5″ Heavy.... r = ·60	Fixed Ends....	11760	8420	6670	5000	3500	2410	1720	1290	980
	Flat Ends......	11760	8420	6260	4350	2900	2070	1460	970	650
	Hinged Ends...	11110	7500	5000	3060	1800	1140	790	580	420
	Round Ends....	10140	5950	3440	1760	1000	650	440	320	230
5″ Light..... r = ·60	Fixed Ends....	11760	8420	6670	5000	3500	2410	1720	1290	980
	Flat Ends......	11760	8420	6260	4350	2900	2070	1460	970	650
	Hinged Ends...	11110	7500	5060	3060	1800	1140	790	580	420
	Round Ends....	10140	5950	3440	1760	1000	650	440	320	230
4″ Heavy.... r = ·62	Fixed Ends....	12010	8730	6910	5310	3830	2660	1870	1440	1070
	Flat Ends......	12010	8730	6600	4730	3190	2260	1650	1140	770
	Hinged Ends...	11380	7830	5390	2460	2020	1270	890	640	470
	Round Ends....	10440	6310	3760	2040	1150	720	510	360	250
4″ Light..... r = ·61	Fixed Ends....	11130	7770	5750	3890	2520	1690	1200	870	.....
	Flat Ends......	11130	7730	5280	3250	2160	1430	880	530	.....
	Hinged Ends...	10420	6590	4060	2060	1200	770	540	350	.....
	Round Ends....	9290	4960	2470	1180	680	430	290	200	.....
3″ Heavy.... r = ·59	Fixed Ends....	11670	8370	6620	4910	3400	2310	1660	1240	940
	Flat Ends......	11670	8370	6210	4280	2830	2830	1390	920	600
	Hinged Ends...	11010	7430	5010	2970	1730	1100	760	560	390
	Round Ends....	10020	5880	3390	1710	960	630	420	300	210
3″ Light..... r = ·53	Fixed Ends....	11310	7900	5960	4130	2730	1810	1320	960	710
	Flat Ends......	11310	7870	5520	3460	2320	1580	1000	630	400
	Hinged Ends...	10620	6780	4300	2220	1310	850	590	410	260
	Round Ends....	9530	5160	2690	1290	740	480	320	220	160

ROLLED ANGLES AS STRUTS.

Tables Nos. 9 and 10 apply to even-legged angles acting as struts. As described in the head notes, the angle is considered free to yield in its weakest direction, that is in the direction of the least radius of gyration.

If the angle is prevented from failing in this direction, by bracing or otherwise, its resistance will be increased to some extent, and a correction can be made by taking the greatest instead of the least radius of gyration into the calculation.

Example.—An angle strut with flat ends, whose dimensions are $4 \times 4 \times \frac{3}{8}$ inches, and 12 feet long, has a least radius of gyration of .81 inch, and greatest radius of gyration 1.24. When the strut has no lateral support the value of $\frac{l}{r}$ would be $\frac{144}{.81} =$ 178. (See table on page 98.) By Table No. 2 the safe load would be 3,580 lbs. per square inch.

If this strut is now braced so that it cannot fail in the weakest direction, that is in the line of a diagonal from the corner of the angle, but is free to fail in the direction of its legs, then the value of $\frac{l}{r}$ becomes $\frac{144}{1.24} = 116$, and the safe load by the tables becomes 6,500 lbs. per square inch.

STRUTS COMPOSED OF SEVERAL ANGLES.

If a strut is composed of several angles, properly braced together, so that the angles cannot fail individually, find the least radius of gyration of the section in the manner described on page 111, and thus the working resistance of the strut from Table No. 2, as described before.

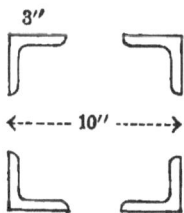

Example. — What is the working resistance of a flat-ended strut 10″ square outside, and 18 feet long, composed of four 3 × 3 angles connected by triangular bracing ?

The radius of gyration as found on page 111, is 4.21 inches. $\frac{l}{r} = 51$.

Safe load per square inch by Table No. 2 = 10,800 lbs.

But the angles will fail individually if the bracing is not sufficient. To determine the greatest distance apart for centres of bracing, consider each angle as a strut bearing 10,800 lbs. per square inch of section. The least radius of gyration for a single angle is .60 inch. By Table No. 2, the value of $\dfrac{l}{r}$ corresponding to the pressure of 10,800 is 51, as found above. Therefore .60 × 51 = 30 inches, which is the greatest distance apart for centres of bracing. For properly designed struts of the foregoing section, the resistance per square inch may be ascertained approximately by means of table No. 18, page 158, although the former kind of column should be somewhat stronger than the latter per unit of section.

STRUTS OF UNEVEN ANGLES.

When uneven angles are used as struts, find the value of $\dfrac{l}{r}$ by means of the least radius of gyration as found on page 99, and the corresponding resistance per square inch of section by table No. 2 as before. If the angle is braced in such a manner that failure cannot occur diagonally, it will then fail in the direction of the shortest leg, and if braced in this direction also, it will be forced to fail in the direction of the longest leg. The resistance in either direction can readily be found by means of the respective radii of gyration, as given in columns VII, VIII, IX, page 99.

It is frequently desirable to use a pair of uneven angles, braced together in the direction of the shortest legs.

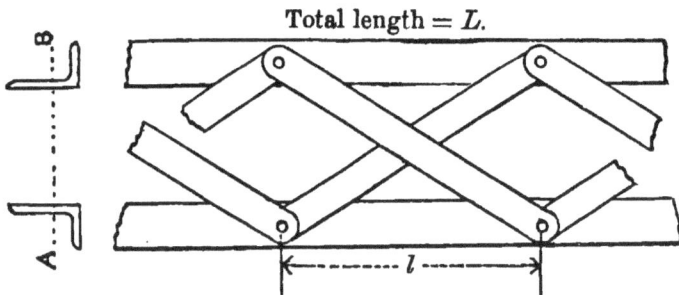

Total length = L.

For this form the least radius of gyration for the combined

sections will be the same as the greatest radius of gyration for a single angle. Therefore take in the tables of elements of uneven angles, the greatest radius, or that corresponding to axis $A B$, when estimating the strength of the combined sections, and the least radius when determining the distance between centres of bracing.

Example.—A flat-ended strut, 16 feet long, is composed of two uneven angles, each $6 \times 4 \times \frac{1}{2}$ inches, and 4.75 square inches sectional area. The angles are braced together in the direction of the short legs. What is the greatest safe load for the strut, and what the greatest distance between centres of bracing measured on the leg of the angle?

By the tables on page 99, the greatest radius of gyration = 1.9 inches, therefore $\frac{l}{r} = 101$.

By Table No. 2 we have for this 7,450 lbs. per square inch, or 70,700 lbs. for the whole strut. The least radius of gyration is .92 inch, which multiplied by 101 gives 92.9 inches as the greatest distance between centres of bracing.

To find the greatest distance apart centres of bracing (l) should be it is only necessary to remember that $\frac{l}{r}$ should not exceed $\frac{L}{R}$.

$l =$ distance between bracing centres.

$r =$ least radius of gyration of single angle.

$L =$ total length of strut.

$R =$ least radius of gyration of combined section.

When struts of any section are hinged, in order to utilize the maximum efficiency of the strut it is of the utmost importance to keep the centre of pin in line with the centre of gravity of cross section of the strut. In the tables of elements 94–101, the positions of centres of gravity are accurately defined.

No. 9.

PENCOYD ANGLES AS STRUTS.

GREATEST SAFE LOAD IN LBS. PER SQUARE INCH OF SECTION USING THE FACTORS OF SAFETY OF PREVIOUS TABLES.

SIZE OF ANGLE.	CONDITION OF ENDS.	LENGTH IN FEET.								
		2	4	6	8	10	12	14	16	18
6" × 6" r = 1·18	Fixed Ends....	14380	11670	9920	8370	7410	6580	5710	4870	4060
	Flat Ends......	14380	11670	9920	8370	7330	6160	5230	4220	3400
	Hinged Ends...	13940	11010	9140	7430	6100	4960	4010	2930	2180
	Round Ends....	13330	10020	7720	5880	4440	3340	2430	1690	1260
5" × 5" r = ·99	Fixed Ends ...	13840	11040	8960	7630	6620	5590	4570	3690	2940
	Flat Ends.....	13840	11040	8960	7590	6210	5080	3920	3070	2480
	Hinged Ends...	13350	10330	8080	6410	5010	3840	2640	1930	1430
	Round Ends....	12670	9170	6580	4760	3390	2310	1530	1090	810
4" × 4" r = ·80	Fixed Ends....	13030	10090	8000	6750	5470	4250	3280	2470	1870
	Flat Ends......	13030	10090	7970	6370	4930	3580	2730	2120	1650
	Hinged Ends...	12460	9320	6910	5170	3680	2320	1650	1170	890
	Round Ends....	11670	7930	5200	3540	2190	1350	920	670	510
3½" × 3½" r = ·69	Fixed Ends....	12520	9270	7370	5920	4520	3310	2410	1790	1400
	Flat Ends......	12520	9270	7270	5470	3870	2760	2070	1550	1090
	Hinged Ends...	11920	8420	6040	4250	2600	1670	1140	830	620
	Round Ends....	11060	6950	4380	2640	1500	930	650	470	350
3" × 3" r = ·58	Fixed Ends....	11760	8420	6670	5000	3500	2410	1720	1290	980
	Flat Ends......	11760	8420	6260	4350	2900	2070	1460	970	650
	Hinged Ends...	11110	7500	5060	3060	1800	1140	790	580	420
	Round Ends....	10140	5950	3440	1760	1000	650	440	320	230
2¾" × 2¾" r = ·54	Fixed Ends....	11400	8090	6170	4370	2940	1940	1440	1040	780
	Flat Ends......	11400	8070	5740	3710	2480	1730	1140	720	460
	Hinged Ends...	10720	7040	4530	2440	1430	930	640	450	300
	Round Ends...	9660	5350	2910	1420	810	540	360	240	170
2½" × 2½" r = ·49	Fixed Ends....	11040	7680	5630	3760	2410	1630	1130	830	....
	Flat Ends.....	11040	7640	5130	3130	2070	1350	830	490	....
	Hinged Ends...	10330	6470	3900	1970	1140	740	510	320	....
	Round Ends....	9170	4830	2350	1120	650	410	270	190	....

No. 9.
PENCOYD ANGLES AS STRUTS.

The radius of gyration is taken about the axis $A\,B$, which also indicates the direction of pin if the strut is hinged.

r in marginal columns indicates radius of gyration around axis $A\,B$.

20	22	24	26	28	30	32	34	36	Condition of Ends.	Size of Angle.
					LENGTH IN FEET.					
3400	2780	2310	1910	1660	1440	1240	1060	940	Fixed Ends...	
2830	2360	2000	1690	1390	1140	920	750	600	Flat Ends......	$6'' \times 6''$
1730	1340	1100	910	760	640	560	460	390	Hinged Ends...	
960	760	630	530	420	360	300	240	210	Round Ends....	
2360	1870	1590	1340	1100	950	810	690		Fixed Ends....	
2030	1650	1310	1020	800	620	470	390		Flat Ends......	$5'' \times 5''$
1120	890	720	600	490	400	310	250		Hinged Ends...	
640	510	400	330	260	220	180	150		Round Ends....	
1540	1230	1000	820	680					Fixed Ends....	
1250	910	670	490	380					Flat Ends......	$4'' \times 4''$
690	550	430	320	240					Hinged Ends...	
380	300	230	190	150					Round Ends....	
1070	870	690							Fixed Ends....	
770	530	390							Flat Ends......	$3\frac{1}{2}'' \times 3\frac{1}{2}''$
470	350	250							Hinged Ends...	
250	200	150							Round Ends....	
740									Fixed Ends....	
430									Flat Ends......	$3'' \times 3''$
280									Hinged Ends...	
170									Round Ends....	
									Fixed Ends....	
									Flat Ends......	$2\frac{3}{4}'' \times 2\frac{3}{4}''$
									Hinged Ends...	
									Round Ends....	
									Fixed Ends....	
									Flat Ends......	$2\frac{1}{2}'' \times 2\frac{1}{2}''$
									Hinged Ends...	
									Round Ends....	

No. 10.

PENCOYD ANGLES AS STRUTS.

GREATEST SAFE LOADS IN LBS. PER SQUARE INCH OF SECTION.

(See remarks at head of Table No. 9.)

SIZE OF ANGLE.	CONDITION OF ENDS.	LENGTH IN FEET.								
		2	4	6	8	10	12	14	16	18
2¼″ × 2¼″ r = ·44	Fixed Ends....	10600	7190	5000	3090	1870	1290	890	640	....
	Flat Ends......	10600	7020	4350	2600	1650	970	550	340	....
	Hinged Ends...	9860	5800	3060	1530	890	580	360	230	....
	Round Ends....	8600	4150	1760	860	510	320	200	140	....
2″ × 2″ r = ·39	Fixed Ends....	10000	6670	4170	2410	1500	980	660	...	....
	Flat Ends......	10000	6260	3500	2070	1200	650	370	....	....
	Hinged Ends...	9230	5060	2250	1140	670	420	230	....	....
	Round Ends....	7820	3440	1310	650	370	230	150	...	....
1¾″ × 1¾″ r = ·35	Fixed Ends....	9430	6090	3500	1870	1160	740	....	....	....
	Flat Ends......	9430	5650	2900	1650	850	430	....	....	....
	Hinged Ends...	8600	4440	1800	890	520	280	....	..	....
	Round Ends....	7140	2820	1000	510	280	170	....	....	....
1½″ × 1½″ r = ·31	Fixed Ends....	8650	5190	2590	1390	810	....	....	...	....
	Flat Ends......	8650	4590	2210	1080	480	....	....	...	....
	Hinged Ends...	7750	3310	1230	620	310	...	....	...	....
	Round Ends....	6220	1940	700	350	180	....	....	...	....
1¼″ × 1¼″ r = ·26	Fixed Ends....	7860	4000	1750	920	....	....	....	...	....
	Flat Ends......	7830	3340	1500	580	....	....	....	...	....
	Hinged Ends...	6720	2130	810	380	....	....	....	...	....
	Round Ends....	5100	1230	450	210	....	...	...	...	....
1″ × 1″ r = ·20	Fixed Ends....	6670	2410	980	....	....	....	....	...	....
	Flat Ends......	6260	2070	650	....	....	....	....	...	....
	Hinged Ends...	5060	1140	420	....	....	....	....	...	....
	Round Ends....	3440	650	230	....	....	....	....	...	..

TEE STRUTS.

The following tables are for even tees. For single uneven tees, find the least radius of gyration from the table of elements, page 101, and proceed as described for angle struts, on page 135.

When a pair of uneven tees are braced together in the direction of the shortest leg, they form a single strut, whose least radius of gyration is the same as the greatest radius of gyration for a single tee.

Therefore, when determining the resistance of the combined strut, take the greatest radius of gyration from the table on page 101, and the least radius of gyration, when determining the distance between centres of lateral bracing.

Example.—A pair of uneven tees 5 × 2¼ inches, whose total area is 6.1 square inches, are braced together in the direction of the shortest leg, forming a single hinged-ended strut 15 feet long. What is the greatest safe load, and what the greatest distance between centres of lateral bracing ?

By table on page 101, greatest radius of gyration = 1.14 inches,

$\dfrac{l}{r}$ = 158, which by Table No. 2 gives 3,100 lbs. per square inch,

or 18,900 lbs. total greatest safe load.

Least radius of gyration = .72, which multiplied by 158 gives 113 inches as the greatest distance between centres of lateral bracing.

No. 11.

PENCOYD TEES AS STRUTS.

GREATEST SAFE LOAD IN LBS. PER SQUARE INCH OF SECTION.

When the strut is free to fail in the direction C. D. Using factors of safety given in previous table.

SIZE OF TEE.	CONDITION OF ENDS.	LENGTH IN FEET.						
		2	4	6	8	10	12	14
4″ × 4″ $r = .84$	Fixed Ends....	13160	10260	8140	6910	5670	4530	3500
	Flat Ends......	13160	10260	8120	6600	5.80	3870	2900
	Hinged Ends...	12610	9500	7100	5390	3950	2600	1800
	Round Ends....	11840	8150	5420	3760	2390	1500	1000
3½″ × 3½″ $r = .74$	Fixed Ends....	12780	9590	7630	6220	4910	3660	2710
	Flat Ends.. ...	12780	9590	7590	5790	4260	3040	2300
	Hinged Ends...	12190	8780	6410	4580	2970	1910	1300
	Round Ends....	11360	7330	4760	2960	1710	1070	740
3″ × 3″ $r = .62$	Fixed Ends....	11890	8650	6830	5190	3690	2590	1820
	Flat Ends......	11890	8650	6490	4590	3070	2210	1590
	Hinged Ends...	11240	7750	5280	3310	1930	1230	860
	Round Ends....	10290	6220	3650	1940	1090	700	490
2½″ × 2½″ $r = .55$	Fixed Ends....	11400	8090	6170	4370	2940	1930	1440
	Flat Ends......	11400	8070	5740	3710	2480	1720	1140
	Hinged Ends...	10720	7040	4530	2440	1430	920	640
	Round Ends....	9660	5350	2910	1420	810	540	360
2¼″ × 2¼″ $r = .47$	Fixed Ends....	10770	7410	5270	3370	2070	1430	990
	Flat Ends......	10770	7330	4680	2800	1830	1120	670
	Hinged Ends...	10040	6100	3410	1710	990	640	420
	Round Ends....	8820	4440	2010	950	570	350	230
2″ × 2″ $r = .43$	Fixed Ends....	10340	6990	4690	2800	1730	1140	790
	Flat Ends......	10340	6720	4040	2380	1470	840	460
	Hinged Ends...	9590	5500	2760	1350	790	510	300
	Round Ends....	8260	3870	1590	760	440	270	170
1¾″ × 1¾″ $r = .37$	Fixed Ends....	9590	6220	3660	1980	1250	800	.....
	Flat Ends......	9590	5790	3040	1760	930	470	.....
	Hinged Ends...	8780	4580	1910	950	560	310	.....
	Round Ends....	.7330	2960	1070	550	300	180	.....
1½″ × 1½″ $r = .32$	Fixed Ends....	8800	5390	2760	1500	880	.....	.....
	Flat Ends......	8800	4630	2340	1200	540	.....	.....
	Hinged Ends...	7910	3570	1330	670	360	.....	.....
	Round Ends....	6400	2120	750	370	200	.....	.....

No. 11.

PENCOYD TEES AS STRUTS.

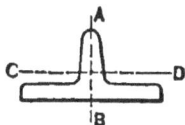

Radius of gyration taken around axis *A. B.* which also indicates the direction of pin when strut is hinged. *r* in marginal columns indicates radius of gyration around axis *A. B.*

16	18	20	22	24	26	28	CONDITION OF ENDS.	SIZE OF TEE.
2660	2020	1650	1350	1070	910	740	Fixed Ends....	$4'' \times 4''$
2260	1790	1380	1030	770	570	430	Flat Ends......	
1270	970	750	600	470	370	280	Hinged Ends...	
720	560	420	330	250	200	170	Round Ends....	$r = .84$
1980	1560	1250	990	600	.....	.....	Fixed Ends....	$3\frac{1}{2}'' \times 3\frac{1}{2}''$
1760	1300	930	670	470	.....	.....	Flat Ends......	
950	710	560	420	310	.....	.....	Hinged Ends...	
550	400	300	230	180	.....	.....	Round Ends....	$r = .74$
1400	1050	810	.....	.....	.....	.....	Fixed Ends ...	$3'' \times 3''$
1090	730	480	.....	.....	.....	.....	Flat Ends......	
620	450	310	.....	.....	.....	.....	Hinged Ends...	
350	240	180	.....	.....	.....	.....	Round Ends....	$r = .62$
1040	780	.....	.....	.....	.....	.....	Fixed Ends....	$2\frac{1}{2}'' \times 2\frac{1}{2}''$
720	450	.....	.....	.....	.....	.....	Flat Ends......	
450	300	.....	.....	.....	.....	.....	Hinged Ends...	
240	170	.....	.....	.....	.....	.....	Round Ends....	$r = .55$
700	.....	.....	.....	.....	.....	.....	Fixed Ends....	$2\frac{1}{4}'' \times 2\frac{1}{4}''$
400	.....	.....	.....	.....	.....	.....	Flat Ends......	
260	.....	.....	.....	.....	.....	.....	Hinged Ends...	
160	.....	.....	.....	.....	.....	.....	Round Ends....	$r = .47$

SIZE OF TEE.	CONDITION OF ENDS.	2	4	6	8	10	12	14
$1\frac{1}{4}'' \times 1\frac{1}{4}''$	Fixed Ends....	8000	4250	1870	1000	.....	.....	.....
	Flat Ends......	7970	3`80	1650	670	.....	.....	.....
	Hinged Ends...	6910	2320	890	430	.....	.....	.....
$r = .27$	Round Ends....	5200	1350	510	230	.....	.....	.....
$1'' \times 1''$	Fixed Ends....	7860	4000	1750	920	.....	.. ..	.....
	Flat Ends......	7830	3340	1500	580	.....	.....	.....
	Hinged Ends...	6720	2130	810	380	.....	.....	.....
$r = .26$	Round Ends....	5100	1230	450	210	.....	.....	.....

No. 12.
LATTICED CHANNEL STRUTS.

GREATEST SAFE LOAD IN LBS. PER SQUARE INCH OF SECTION,
USING FACTORS OF SAFETY GIVEN IN PREVIOUS TABLES.

For a pair of braced channels or for a single channel secured from flexure in the direction of the flanges and liable to fail only in the direction of the web $C D$.

r in the marginal columns gives the radius of gyration for axis $A B$, or for either axis of the combined pair of channels. See description, page 121.

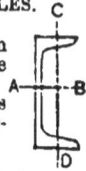

SIZE OF CHANNEL	CONDITION OF ENDS.	LENGTH IN FEET.								
		6	8	10	12	14	16	18	20	22
15″ r — 5.51 D = 12.7 d = 8.9	Fixed Ends....	.....	.....	14110	13570	12900	12400	11890	11400	11040
	Flat Ends......	.....	.....	14110	13570	12900	12400	11890	11400	11040
	Hinged Ends...	.....	.....	13640	13050	12320	11780	11240	10720	10330
	Round Ends....	.....	.....	13000	12330	11520	10900	10290	9660	9170
				2.05	2.46	2.87	3.28	3.69	4.10	4.51
12″H'y r — 4.55 D = 10.3 d = 7.6	Fixed Ends....	.....	14240	13570	12780	12140	11580	11130	10600	10170
	Flat Ends	.....	14240	13570	12780	12140	11580	11130	10600	10170
	Hinged Ends...	.....	13790	13050	12190	11510	10910	10420	9860	9410
	Round Ends....	.....	13160	12330	11360	10600	9000	9290	8600	8040
			1.61	2.02	2.42	2.83	3.23	3.64	4.04	4.44
12″L't r — 4.55 D = 10.3 d = 7.7	Fixed Ends....	.....	14240	13570	12780	12140	11580	11130	10600	10170
	Flat Ends....	.....	14240	13570	12780	12140	11580	11130	10600	10170
	Hinged Ends...	.....	13790	13350	12190	11510	10910	10420	9860	9410
	Round Ends....	.....	13160	12330	11360	10600	9900	9290	8600	8040
			1.30	1.62	1.94	2.27	2.59	2.92	3.24	3.66
10″H'y r — 3.92 D = 9.0 d = 6.3	Fixed Ends....	.....	13840	12900	12140	11490	10950	10430	9920	9430
	Flat Ends	.....	13840	12900	12140	11490	10950	10430	9920	9430
	Hinged Ends...	.....	13350	12320	11510	10820	10230	9680	9140	8600
	Round Ends....	.....	12670	11520	10600	9780	9050	8370	7720	7140
			1.71	2.14	2.57	2.99	3.42	3.85	4.28	4.71
10″L't r — 3.89 D = 8.9 d = 6.3	Fixed Ends....	.....	13700	12900	12140	11490	10950	10340	9840	9350
	Flat Ends......	.....	13700	12900	12140	11490	10950	10340	9840	9350
	Hinged Ends...	.....	13200	12330	11510	10820	10230	9590	9050	8510
	Round Ends....	.....	12500	11520	10600	9780	9050	8260	7630	7040
			1.42	1.77	2.13	2.48	2.84	3.19	3.55	3.90
9″He'vy r — 3.46 D = 8.1 d = 5.4	Fixed Ends....	14240	13300	12400	11580	10950	10340	9760	9190	8650
	Flat Ends......	14240	13300	12400	11580	10950	10340	9760	9190	8650
	Hinged Ends...	13790	12760	11780	10910	10230	9590	8960	83 0	7750
	Round Ends....	13160	12000	10900	9900	9050	8260	7530	6850	6220
		1.18	1.53	1.97	2.36	2.76	3.15	3.55	3.94	4.33
9″Light r — 3.42 D = 7.9 d = 5.6	Fixed Ends....	14240	13300	12400	11580	10950	10340	9760	9190	8650
	Flat Ends......	14240	13300	12400	11580	10950	10340	9760	9110	8650
	Hinged Ends...	13790	12760	11780	10910	10230	9590	8960	8330	7750
	Round Ends....	13160	12000	10900	9900	9050	8260	7530	6850	6220
		1.03	1.38	1.72	2.08	2.41	2.75	3.10	3.44	3.78

No. 12.
LATTICED CHANNEL STRUTS.

GREATEST SAFE LOAD IN LBS. PER SQUARE INCH OF SECTION,
USING FACTORS OF SAFETY GIVEN IN PREVIOUS TABLES.

The channels must be connected so as to insure unity of action and separated not less than the distances D or d respectively, given in inches in the marginal columns. Figures in heavy type under each length represent the greatest distances apart in feet on each channel that centres of lateral bracing should be placed.

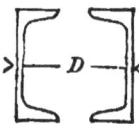

24	26	28	30	32	34	36	38	40	CONDITION OF ENDS.	SIZE OF CHANNEL.
10690	10260	9920	9590	9190	8880	8580	8280	8090	Fixed Ends....	15"
10690	10260	9920	9590	9190	8880	8580	8270	8070	Flat Ends......	r = 5.51
9950	9500	9140	8780	8330	8000	7670	7300	7040	Hinged Ends..	D = 12.7
8710	8150	7720	7330	6850	6490	6130	5730	5350	Round Ends ...	d = 8.9
4.92	5.33	5.74	6.15	6.56	6.97	7.38	7.79	8.20		
9760	9270	8880	8500	8230	7950	7720	7500	7240	Fixed Ends....	12"H'y
9760	9270	8880	8500	8220	7920	7680	7450	7080	Flat Ends......	r = 4.55
8960	8420	8000	7580	7240	6840	6530	6220	5860	Hinged Ends..	D = 10.2
7530	6950	6490	6040	5600	5230	4890	4560	4210	Round Ends....	d = 7.6
4.85	5.25	5.66	6.06	6.47	6.87	7.28	7.68	8.09		
9760	9270	8880	8500	8230	7950	7720	7500	7280	Fixed Ends....	12"L't
9760	9270	8880	8500	8220	7920	7680	7450	7140	Flat Ends......	r = 4.55
8960	8420	8000	7580	7240	6840	6530	6220	5920	Hinged Ends...	D = 10.2
7530	6950	6490	6040	5660	5230	4890	4560	4270	Round Ends....	d = 7.7
3.89	4.21	4.54	4.86	5.19	5.51	5.84	6.16	6.49		
8960	8420	8140	7860	7590	7320	7070	6830	6580	Fixed Ends....	10"H'y
8960	8420	8120	7830	7540	7210	6840	6490	6160	Flat Ends......	r = 3.92
8080	7500	7100	6720	6340	5980	5620	5280	4960	Hinged Ends..	D = 9.0
6580	5950	5420	5100	4690	4320	3980	3650	3340	Round Ends....	d = 6.2
5.13	5.56	5.99	6.42	6.85	7.28	7.71	8.14	8.57		
8880	8420	8140	7810	7540	7280	7030	6790	6530	Fixed Ends....	10"L't
8880	8420	8120	7780	7500	7140	6780	6430	6110	Flat Ends......	r = 3.89
8000	7500	7100	6650	6280	5920	5560	5220	4910	Hinged Ends..	D = 8.9
6490	5950	5420	5030	4630	4270	3920	3600	3290	Round Ends...	d = 6.2
4.26	4.61	4.97	5.32	5.68	6.03	6.39	6.74	7.10		
8280	7950	7630	7320	7030	6750	6440	6130	5840	Fixed Ends...	9"He'vy
8270	7920	7590	7210	6780	6370	6020	5700	5380	Flat Ends......	r = 3.45
7300	6840	6410	5980	5560	5170	4820	4480	4160	Hinged Ends.	D = 8.1
5730	5230	4760	4320	3920	3540	3200	2870	2550	Round Ends....	d = 5.4
4.73	5.12	5.52	5.91	6.30	6.70	7.09	7.49	7.88		
8230	7900	7590	7280	6990	6710	6400	6090	5800	Fixed Ends....	9"Light.
8220	7870	7540	7140	6720	6310	5970	5650	5340	Flat Ends......	r = 3.43
7240	6780	6340	5920	5500	5110	4770	4440	4120	Hinged Ends...	D = 7.9
5660	5160	4690	4270	3870	3490	3150	2820	2510	Round Ends...	d = 5.5
4.13	4.47	4.82	5.16	5.50	5.85	6.19	6.54	6.88		

11

No. 13.
LATTICED CHANNEL STRUTS.

GREATEST SAFE LOAD IN LBS. PER SQUARE INCH OF SECTION, USING FACTORS OF SAFETY GIVEN IN PREVIOUS TABLES.

For a pair of braced channels or for a single channel secured from flexure in the direction of the flanges and liable to fail only in the direction of the web $C\,D$.

r in the marginal columns gives the radius of gyration for axis $A\,B$, or for either axis of the combined pair of channels. See description, page 121.

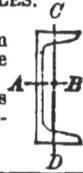

SIZE OF CHANNEL	CONDITION OF ENDS.	LENGTH IN FEET.								
		4	6	8	10	12	14	16	18	20
8″ He'vy r — 3·06 D = 7·3 d — 4·8	Fixed Ends		13840	12900	11890	11130	10430	9760	9190	8580
	Flat Ends		13840	12900	11890	11130	10430	9760	9190	8580
	Hinged Ends		13350	12320	11240	10420	9680	8960	8330	7670
	Round Ends		12670	11520	10290	9290	8370	7530	6850	6130
			1.39	1.86	2.32	2.78	3.25	3.71	4.18	4.64
8″ Light r — 3·09 D = 7·1 d — 5·0	Fixed Ends		13970	12900	11890	11130	10520	9840	9190	8580
	Flat Ends		13970	12900	11890	11130	10520	9840	9190	8580
	Hinged Ends		13500	12320	11240	10420	9770	9050	8330	7670
	Round Ends		12830	11520	10290	9290	8490	7630	6850	6130
			1.16	1.55	1.94	2.33	2.72	3.10	3.49	3.88
7″ He'vy r — 2·68 D = 6·4 d — 3·9	Fixed Ends		13430	12270	11310	10520	9760	9040	8370	7950
	Flat Ends		13430	12270	11310	10520	9760	9040	8370	7920
	Hinged Ends		12900	11650	10620	9770	8960	8160	7430	6840
	Round Ends		12170	10750	9530	8490	7530	6670	5890	5230
			1.46	1.95	2.43	2.91	3.40	3.88	4.37	4.85
7″ Light r — 2·64 D — 6·1 d — 4·2	Fixed Ends		13430	12270	11310	10430	9680	8960	8330	7900
	Flat Ends		13430	12270	11310	10430	9680	8960	8320	7870
	Hinged Ends		12900	11650	10620	9680	8870	8080	7370	6780
	Round Ends		12170	10750	9530	8370	7430	6580	5810	5160
			1.32	1.76	2.20	2.64	3.03	3.52	3.96	4.40
6″ He'vy r — 2·36 D — 5·8 d — 3·3	Fixed Ends	14380	12900	11670	10770	9920	9110	8370	7860	7410
	Flat Ends	14380	12900	11670	10770	9920	9110	8370	7890	7330
	Hinged Ends	13940	12320	11010	10040	9140	8250	7430	6720	6100
	Round Ends	13330	11520	10020	8820	7720	6760	5880	5100	4440
		1.14	1.70	2.27	2.84	3.41	3.98	4.54	5.11	5.68
6″ Light r — 2·27 D — 5·3 d — 3·4	Fixed Ends	14240	12900	11580	10690	9760	8880	8180	7720	7240
	Flat Ends	14240	12900	11580	10690	9760	8880	8170	7680	7080
	Hinged Ends	13790	12320	10910	9950	8960	8000	7170	6530	5860
	Round Ends	13160	11520	9900	8710	7530	6490	5490	4890	4210
		.90	1.35	1.80	2.25	2.70	3.15	3.60	4.05	4.50
5″ He'vy r — 1·93 D — 4·9 d — 2·5	Fixed Ends	18700	12140	10860	9840	8800	8090	7500	6990	6490
	Flat Ends	18700	12140	10860	9840	8800	8070	7450	6720	6070
	Hinged Ends	13200	11510	10130	9050	7910	7040	6220	5500	4860
	Round Ends	12500	10600	8930	7630	6400	5350	4560	3870	3240
		1.18	1.74	2.32	2.90	3.48	4.06	4.64	5.22	5.80

No. 13.
LATTICED CHANNEL STRUTS.

GREATEST SAFE LOAD IN LBS. PER SQUARE INCH OF SECTION,
USING FACTORS OF SAFETY GIVEN IN PREVIOUS TABLES.

The channels must be connected so as to insure unity of action and separated not less than the distances D or d respectively, given in inches in the marginal columns. Figures in heavy type under each length represent the greatest distances apart in feet on each channel that centres of lateral bracing should be placed.

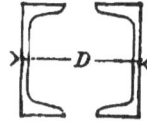

22	24	26	28	30	32	34	36	38	CONDITION OF ENDS.	SIZE OF CHANNEL
8140	7770	7410	7070	6750	6440	6090	5750	5430	Fixed Ends....	8″ He'vy
8120	7730	7330	6840	6370	6020	5650	5280	4880	Flat Ends	r = 3·06
7100	6590	6100	5620	5170	4820	4440	4060	3620	Hinged Ends...	D = 7·2
5420	4960	4440	3980	3540	3200	2820	2470	2150	Round Ends ...	d = 4·8
5.10	5.57	6.03	6.50	6.96	7.42	7.89	8.35	8.82		
8140	7810	7450	7110	6790	6490	6130	5800	5510	Fixed Ends....	8″ Light.
8120	7780	7390	6900	6430	6070	5700	5340	4980	Flat Ends......	r = 3·09
7100	6650	6160	5680	5220	4860	4480	4120	3730	Hinged Ends...	D = 7·1
5420	5020	4500	4030	3600	3240	2870	2510	2230	Round Ends ...	d = 5·0
4.27	4.66	5.04	5.43	5.62	6.21	6.60	6.99	7.38		
7540	7190	6830	6440	6050	5670	5310	4950	4570	Fixed Ends ..	7″ He'vy
7500	7020	6490	6020	5610	5180	4730	4300	3920	Flat Ends......	r = 2·68
6280	5800	5280	4820	4390	3950	3460	3010	2640	Hinged Ends...	D = 6·5
4630	4150	3650	3200	2760	2390	2040	1730	1530	Round Ends ...	d = 3·9
5.34	5.82	6.31	6.79	7.28	7.76	8.25	8.73	9.22		
7500	7110	6750	6350	5960	5590	5190	4820	4450	Fixed Ends....	7″ Light.
7450	6900	6370	5930	5520	5080	4590	4170	3790	Flat Ends......	r = 2·64
6220	5680	5170	4720	4300	3840	3310	2890	2520	Hinged Ends...	D = 6·1
4560	4030	3540	3100	2690	2310	1940	1660	1460	Round Ends ...	d = 4·2
4.84	5.28	5.72	6.16	6.59	7.03	7.47	7.91	8.35		
6990	6580	6130	5710	5270	4870	4450	4060	3730	Fixed Ends....	6″ He'vy
6720	6160	5700	5230	4680	4220	3790	3400	3100	Flat Ends......	r = 2·36
5500	4960	4480	4010	3410	2930	2520	2180	1950	Hinged Ends...	D = 5·6
3870	3340	2870	2430	2010	1690	1460	1260	1100	Round Ends ...	d = 3·3
6.25	6.82	7.38	7.95	8.52	9.09	9.66	10.22	10.79		
6830	6350	5920	5470	5030	4610	4170	3830	3400	Fixed Ends....	6″ Light.
64·0	5930	5470	4930	4390	3960	3500	3190	2870	Flat Ends......	r = 2·27
5280	4720	4250	3680	3110	2640	2250	2020	1770	Hinged Ends...	D = 5·5
3650	3100	2640	2190	1790	1550	1310	1150	980	Round Ends ...	d = 3·5
4.95	5.40	5.85	6.30	6.75	7.20	7.65	8.10	8.55		
5920	5430	4910	4410	3930	3530	3150	2780	2500	Fixed Ends...	5″ He'vy
5470	4880	4260	3750	3280	2920	2640	2360	2140	Flat Ends......	r = 1·93
4250	3620	2970	2480	2080	1820	1570	1340	1180	Hinged Ends...	D = 4·9
2640	2150	1710	1440	1190	1010	880	760	670	Round Ends ...	d = 2·5
6.38	6.96	7.54	8.12	8.70	9.28	9.86	10.44	11.02		

No. 14.
LATTICED CHANNEL STRUTS.

GREATEST SAFE LOAD IN LBS. PER SQUARE INCH OF SECTION, USING FACTORS OF SAFETY GIVEN IN PREVIOUS TABLES.

For a pair of braced channels or for a single channel secured from flexure in the direction of the flanges and liable to fail only in the direction of the web $C\,D$.

r in the marginal columns gives the radius of gyration for axis $A\,B$, or for either axis of the combined pair of channels. See description, page 121.

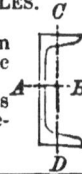

SIZE OF CHANNEL.	CONDITION OF ENDS.	LENGTH IN FEET.								
		2	4	6	8	10	12	14	16	18
5″ Light, r = 1·89, D = 4·6, d = 2·8	Fixed Ends		13570	12010	10770	9680	8650	8000	7410	6870
	Flat Ends		13570	12010	10770	9680	8650	7970	7330	6550
	Hinged Ends		13050	11380	10040	8870	7750	6910	6100	5340
	Round Ends		12330	10440	8820	7430	6220	5200	4440	3710
			.96	1.43	1.91	2.39	2.87	3.35	3.83	4.31
4″ Heavy, r = 1·55, D = 4·0, d = 1·9	Fixed Ends		12900	11220	9840	8650	7810	7150	6490	5840
	Flat Ends		12900	11220	9840	8650	7780	6960	6070	5380
	Hinged Ends		12320	10520	9050	7750	6650	5740	4860	4160
	Round Ends		11520	9410	7630	6220	5030	4090	3240	2550
			1.29	1.94	2.58	3.23	3.88	4.52	5.17	5.81
4″ Light, r = 1·54, D = 3·8, d = 2·0	Fixed Ends		12900	11130	9840	8580	7810	7110	6440	5800
	Flat Ends		12900	11130	9840	8580	7780	6900	6020	5340
	Hinged Ends		12320	10420	9050	7670	6650	5680	4820	4120
	Round Ends		11520	9290	7630	6130	5030	4030	3200	2510
			1.25	1.87	2.50	3.12	3.74	4.37	4.99	5.62
3″ r = 1·16, D = 3·1, d = 1·1	Fixed Ends	14240	11670	9840	8280	7370	6490	5590	4780	3960
	Flat Ends	14240	11670	9840	8270	7270	6070	5080	4130	3310
	Hinged Ends	13790	11010	9050	7300	6040	4860	3840	2850	2110
	Round Ends	13160	10020	7630	5730	4380	3240	2310	1640	1210
		.79	1.59	2.38	3.18	3.97	4.36	4.76	5.15	5.55
2¼″ r = ·85, D = 2·4, d = ·54	Fixed Ends	13300	10340	8180	6950	5750	4610	3560	2730	2090
	Flat Ends	13300	10340	8170	6660	5290	3960	2950	2320	1840
	Hinged Ends	12760	9590	7170	5450	4060	2980	1840	1310	1000
	Round Ends	12000	8260	5490	3810	2470	1550	1030	740	580
		1.01	2.02	3.03	4.05	5.06	6.07	7.08	8.10	9.11
2″ r = ·74, D = 2·1, d = ·60	Fixed Ends	12780	9590	7630	6220	4910	3690	2710	1980	1580
	Flat Ends	12780	9590	7590	5790	4260	3070	2300	1760	1300
	Hinged Ends	12190	8780	6410	4580	2970	1930	1300	950	710
	Round Ends	11360	7330	4760	2960	1710	1090	740	550	400
		.84	1.68	2.52	3.35	4.19	5.03	5.87	6.70	7.54

No. 14.
LATTICED CHANNEL STRUTS.

GREATEST SAFE LOAD IN LBS. PER SQUARE INCH OF SECTION, USING FACTORS OF SAFETY GIVEN IN PREVIOUS TABLES.

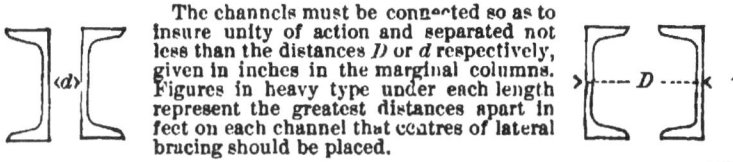

The channels must be connected so as to insure unity of action and separated not less than the distances D or d respectively, given in inches in the marginal columns. Figures in heavy type under each length represent the greatest distances apart in feet on each channel that centres of lateral bracing should be placed.

20	22	24	26	28	30	32	34	36	CONDITION OF ENDS.	SIZE OF CHANNEL.
6310	5800	5270	4740	4210	3790	3370	2970	2640	Fixed Ends....	**5″**
5880	5340	4680	4090	3540	3160	2800	2500	2250	Flat Ends......	Light.
4670	4120	3410	2800	2280	1990	1710	1450	1260	Hinged Ends...	r = 1·88
3050	2510	2010	1620	1330	1130	950	820	720	Round Ends....	D = 4·8
4.73	5.26	5.74	6.22	6.70	7.18	7.66	8.14	8.61		d = 2·8
5190	4570	3960	3460	2970	2590	2240	1920	1740	Fixed Ends....	**4″**
4590	3920	3310	2870	2500	2210	1950	1700	1490	Flat Ends......	Heavy.
3310	2640	2110	1770	1450	1230	1070	910	800	Hinged Ends...	r = 1·55
1940	1530	1210	980	820	700	610	530	450	Round Ends...	D = 4·0
6.46	7.10	7.75	8.39	9.04	9.66	10.32	10.97	11.61		d = 1·9
5150	4490	3930	3400	2940	2540	2200	1900	1700	Fixed Ends....	**4″**
4540	3830	3280	2830	2480	2170	1920	1680	1440	Flat Ends......	Light.
3260	2560	2080	1730	1430	1210	1050	900	780	Hinged Ends...	r = 1·54
1900	14·0	1190	960	810	690	600	520	430	Round Ends....	D = 3·6
6.24	6.86	7.48	8.11	8.73	9.35	9.97	10.60	11.22		d = 2·0
3280	2680	2220	1850	1610	1390	1180	1020	900	Fixed Ends....	**3″**
2730	2280	1940	1620	1330	1080	870	700	560	Flat Ends	
1650	1280	1060	870	730	620	530	440	370	Hinged Ends...	r = 1·16
920	730	610	500	410	350	280	230	200	Round Ends ...	D = 3·1
7.94	8.33	8.72	9.12	9.51	11.90	12.29	12.69	13.08		d = 1·1
1690	1380	1100	930	770	.....	.....	.....	.....	Fixed Ends....	**2¼″**
1430	1060	800	600	450	.....	.....	.....	.....	Flat Ends......	
770	610	490	390	290	.....	.....	.....	.....	Hinged Ends...	r = ·85
430	340	260	210	170	.....	.....	.....	.....	Round Ends....	D = 2·4
10.12	11.13	12.14	13.15	14.17	.....	.....	.....	.....		d = ·64
1250	990	800	.....	.....	.....	.....	.....	.....	Fixed Ends....	**2″**
930	670	470	.....	.....	.....	.....	.....	.....	Flat Ends......	
560	420	310	.....	.....	.....	.....	.....	.....	Hinged Ends...	r = ·74
300	230	180	.....	.....	.....	.....	.....	.....	Round Ends....	D = 2·1
8.38	9.22	10.05	.....	.....	.....	.....	.....	.....		d = ·60

No. 15.

PENCOYD CHANNELS AS STRUTS.

GREATEST SAFE LOADS IN LBS. PER SQ. INCH OF SECTION, WHEN THE
STRUTS ARE FREE TO BEND AT RIGHT ANGLES TO THE WEB OR IN
THE WEAKEST DIRECTION, USING FACTORS OF SAFETY GIVEN IN
PREVIOUS TABLES.

SIZE OF CHANNEL	CONDITION OF ENDS.	LENGTH IN FEET.								
		2	4	6	8	10	12	14	16	18
15″	Fixed Ends....	14240	11580	9680	8180	7240	6350	5430	4570	3790
	Flat Ends......	14240	11580	9680	8170	7080	5930	4880	3920	3160
	Hinged Ends...	13790	10910	8870	7170	5860	4720	3620	2640	1990
r = 1·13	Round Ends....	13160	9900	7430	5490	4210	3100	2150	1530	1130
12″	Fixed Ends....	13570	10690	8580	7320	6220	5110	4060	3220	2520
	Flat Ends......	13570	10690	8580	7210	5790	4490	3400	2690	2160
Heavy....	Hinged Ends...	13050	9950	7670	5980	4580	3210	2180	1610	1200
r = ·92	Round Ends....	12330	8710	6130	4320	2960	1860	1260	900	680
12″	Fixed Ends....	12780	9590	7630	6220	4910	3660	2710	1980	1580
	Flat Ends......	12780	9590	7590	5790	4260	3040	2300	1760	1300
Light.....	Hinged Ends...	12190	8760	6410	4580	2970	1910	1300	950	710
r = ·74	Round Ends....	11360	7330	4760	2960	1710	1070	740	550	400
10″	Fixed Ends...	13160	10260	8140	6910	5670	4530	3500	2660	2020
	Flat Ends......	13160	10260	8120	6600	5190	3870	2900	2260	1790
Heavy....	Hinged Ends...	12610	9500	7100	5390	3950	2600	1800	1270	970
r = ·84	Round Ends....	11840	8150	5420	3760	2390	1500	1000	720	560
10″	Fixed Ends....	12400	9190	7320	5840	4410	3220	2340	1740	1360
	Flat Ends......	12400	9190	7210	5380	3750	2690	2020	1490	1040
Light.....	Hinged Ends...	11780	8330	5980	4160	2480	1610	1110	800	600
r = ·69	Round Ends....	10900	6850	4320	2550	1440	900	630	450	340
9″	Fixed Ends....	12400	9110	7240	5750	4330	3120	2240	1690	1320
	Flat Ends......	12400	9110	7080	5280	3660	2620	1950	1430	1000
Heavy....	Hinged Ends...	11780	8250	5860	4060	2400	1550	1070	770	590
r = ·69	Round Ends....	10900	6760	4210	2470	1390	870	610	430	320
9″	Fixed Ends....	11670	8370	6620	4910	3400	2310	1660	1240	940
	Flat Ends......	11670	8370	6210	4260	2830	2000	1390	920	600
Light.....	Hinged Ends...	11010	7430	5010	2970	1730	1100	760	560	390
r = ·69	Round Ends....	10020	5880	3390	1710	960	630	420	300	210

No. 15.

PENCOYD CHANNELS AS STRUTS.

A----⌐‾‾‾‾‾⌐----B   (with C above center and D below)

r in marginal columns is the radius of gyration taken around axis *A B*.
When strut is hinged the pins are supposed to lie in the direction *A B*.
When the pins are in the direction *C D*, consider the strut flat ended by this table.

LENGTH IN FEET.									CONDITION OF ENDS.	SIZE OF CHANNEL.
20	22	24	26	28	30	32	34	36		
3120	2540	2070	1760	1520	1300	1090	970	840	Fixed Ends....	15"
2620	2170	1830	1520	1220	980	800	640	500	Flat Ends......	
1550	1210	990	820	680	580	490	410	330	Hinged Ends...	
870	690	570	460	370	320	260	220	190	Round Ends ...	r = 1.13
1940	1640	1360	1100	950	790				Fixed Ends....	12"
1730	1360	1040	800	610	460				Flat Ends......	
930	740	600	490	400	300				Hinged Ends...	Heavy.
540	410	340	260	220	170				Round Ends...	r = .92
1250	990	800							Fixed Ends....	12"
930	670	470							Flat Ends......	
560	420	310							Hinged Ends...	Light.
300	230	180							Round Ends....	r = .74
1650	1350	1070	910	740					Fixed Ends....	10"
1380	1030	770	570	430					Flat Ends......	
750	600	470	370	280					Hinged Ends...	Heavy.
420	330	250	200	170					Round Ends....	r = .84
1050	830	670							Fixed Ends....	10"
730	490	370							Flat Ends......	
450	330	240							Hinged Ends...	Light.
240	190	150							Round Ends....	r = .69
1020	810								Fixed Ends....	9"
690	470								Flat Ends......	
440	310								Hinged Ends...	Heavy.
230	180								Round Ends....	r = .68
710									Fixed Ends....	9"
400									Flat Ends. . ..	
260									Hinged Ends...	Light.
160									Round Ends....	r = .59

No. 16.

PENCOYD CHANNELS AS STRUTS.

GREATEST SAFE LOAD IN LBS. PER SQ. INCH OF SECTION WHEN THE STRUTS ARE FREE TO BEND AT RIGHT ANGLES TO THE WEB OR IN THE WEAKEST DIRECTION, USING FACTORS OF SAFETY GIVEN IN PREVIOUS TABLES.

SIZE OF CHANNEL.	CONDITION OF ENDS.	LENGTH IN FEET.								
		2	4	6	8	10	12	14	16	18
8″ Heavy.... r = ·71	Fixed Ends....	12520	9350	7450	6010	4610	3400	2470	1840	1450
	Flat Ends......	12520	9350	7390	5560	3960	2830	2120	1610	1150
	Hinged Ends...	11920	8510	6160	4350	2680	1730	1170	870	650
	Round Ends ...	11060	7040	4500	2730	1550	960	670	500	360
8″ Light .... r = ·60	Fixed Ends....	11760	8420	6670	5000	3500	2410	1720	1290	980
	Flat Ends......	11760	8420	6160	4350	2900	2070	1460	970	650
	Hinged Ends...	11110	7500	5060	3060	1800	1140	790	580	420
	Round Ends ...	10140	5950	3440	1760	1000	650	440	320	230
7″ Heavy.... r = ·65	Fixed Ends....	12140	8880	7030	5470	4000	2830	2000	1550	1170
	Flat Ends......	12140	8880	6780	4930	3340	2400	1780	1260	860
	Hinged Ends...	11510	8000	5560	3680	2130	1370	960	700	530
	Round Ends...	10600	6490	3920	2190	1230	770	560	390	280
7″ Light..... r = ·58	Fixed Ends....	11670	8280	6490	4780	3280	2220	1610	1180	900
	Flat Ends......	11670	8270	6070	4130	2730	1940	1330	870	560
	Hinged Ends...	11010	7300	4860	2850	1650	1060	730	530	370
	Round Ends ...	10020	5730	3240	1640	920	610	410	280	200
6″ Heavy.... r = ·67	Fixed Ends....	12270	9040	7190	5670	4210	3030	2150	1640	1270
	Flat Ends......	12270	9040	7020	5180	3540	2550	1890	1360	950
	Hinged Ends...	11650	8160	5800	3950	2280	1490	1030	740	570
	Round Ends ...	10750	6670	4150	2390	1330	840	590	410	310
6″ Light .... r = ·51	Fixed Ends....	11130	7770	5750	3890	2520	1690	1200	870	....
	Flat Ends......	11130	7730	5280	3250	2160	1430	880	530	....
	Hinged Ends...	10420	6590	4060	2060	1200	770	540	350	...
	Round Ends ...	9290	4960	2470	1180	680	430	290	200	....
5″ Heavy.... r = ·56	Fixed Ends....	11490	8140	6260	4530	3060	2020	1500	1070	820
	Flat Ends......	11490	8120	5830	3870	2570	1790	1200	770	480
	Hinged Ends...	10820	7100	4620	2600	1510	970	670	470	320
	Round Ends ...	9780	5420	3000	1500	850	560	370	250	180

No. 16.

PENCOYD CHANNELS AS STRUTS.

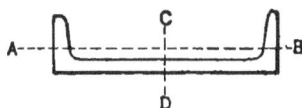

r, in marginal columns, is the radius of gyration taken around axis $A\ B$.
When strut is hinged, the pins are supposed to lie in the direction $A\ B$.
When the pins are in the direction $C\ D$, consider the strut flat ended by this
table.

SIZE OF CHANNEL	CONDITION OF ENDS.	LENGTH IN FEET.								
		2	4	6	8	10	12	14	16	18
5″ Light .... $r = .45$	Fixed Ends....	10600	7190	5000	3090	1870	1290	890	.....	.....
	Flat Ends......	10600	7020	4350	2600	1650	970	550	.....	.....
	Hinged Ends...	9860	5800	3060	1530	890	580	360	.....	.....
	Round Ends ...	8500	4150	1760	860	510	320	200	.....	.....
4″ Heavy.... $r = .50$	Fixed Ends ...	11040	7680	5630	3760	2410	1630	1130	830	....
	Flat Ends......	11040	7640	5130	3130	2070	1350	830	490	.....
	Hinged Ends...	10330	6470	3900	1970	1140	740	510	320	.....
	Round Ends ...	9170	4830	2350	1120	650	410	270	190	.....
4″ Light .... $r = .48$	Fixed Ends....	10860	7500	5390	3500	2180	1500	1040	740	.....
	Flat Ends	10860	7450	4830	2900	1910	1200	720	430	.....
	Hinged Ends...	10130	6220	3570	1800	1040	670	450	280	.....
	Round Ends ...	8930	4560	2120	1000	600	370	240	170	.....
3″ $r = .46$	Fixed Ends....	10690	7320	5110	3220	1940	1360	950	670	.....
	Flat Ends......	10690	7210	4490	2690	1730	1040	610	370	.....
	Hinged Ends ..	9950	5980	3210	1610	930	600	400	240	.....
	Round Ends ...	8710	4320	1860	900	540	340	220	150	.....
2¼″ $r = .42$	Fixed Ends....	10340	6990	4690	2800	1730	1140	790	.....	.....
	Flat Ends......	10340	6720	4040	2380	1470	840	460	. ....	.....
	Hinged Ends...	9590	5500	2760	1350	790	510	300	.....	.....
	Round Ends ...	8260	3870	1590	760	440	270	170	.....	.....
2″ $r = .31$	Fixed Ends....	8650	5190	2~90	1390	810	.....	.....	.....	.....
	Flat Ends......	8650	4590	2210	1080	480	.....	.....	.....	.....
	Hinged Ends...	7750	3310	1230	620	310	.....	.....	.....	.....
	Round Ends ...	6220	1940	700	350	180	.....	.....	.....	.....

WROUGHT IRON COLUMNS OR PILLARS OF ROUND AND SQUARE CROSS SECTION.

Experiments on columns of this class are not very complete, especially as denoting the comparative values for the various end conditions. The following tables, Nos. 17 and 18, are derived partly from experiment on actual columns, extended and completed by comparison with the experiments on rolled struts from which all our previous tables of strut resistances are derived.

Table No. 2 is taken as the basis for the working values. On account of the more perfect symmetry of form possessed by round and square sections than the shapes for which table No. 2 was especially calculated, the safe loads per square inch of section are increased ten (10) per cent. for round columns, and five (5) per cent. for square columns. That is, the factors of safety previously given remaining the same, the ultimate strength is supposed to be 10 and 5 per cent. respectively greater than the rolled struts.

The tables are calculated for certain thicknesses of iron varying from $\frac{1}{4}''$ for $2''$ diameter up to $\frac{3}{4}''$ for $12''$ diameter, as marked in the margins. At the same place R represents the radius of gyration for the diameter and thickness given. When the thickness varies but a little from that given, the strength per square inch of section can be accepted as practically unchanged. But when the variation becomes of importance, the radius of gyration corresponding to the altered thickness will have to be obtained, and the strength of the column then ascertained from table No. 2, as heretofore described.

The following table gives the values of the radius of gyration for round and square columns from 2 to 12 inches diameter, and from $\frac{1}{16}$ of an inch to 1 inch thick.

Example for Round Column :

What is the greatest safe load for a flat-ended round column 6 inches outer diameter, $\frac{1}{2}''$ thick, 8.64 sq. in. area, and 18 feet long. $r = 1.95 \ \frac{l}{r} = 111$. By table No. 2 the corresponding safe load = 6780 lbs. + 10 per cent. = 7460 lbs. per sq. inch of section, or 64,440 lbs. for the column.

For a square column add 5 per cent. to table No. 2, instead of 10 per cent. as above.

RADII OF GYRATION FOR ROUND COLUMNS.

OUTSIDE DIAMETER OF COLUMN IN INCHES.	THICKNESS IN INCHES VARYING BY TENTHS.									
	.1	.2	.3	.4	.5	.6	.7	.8	.9	1.0
	CORRESPONDING RADIUS OF GYRATION IN INCHES.									
2	.67	.64	.61	.58	.56	.54	.52	.51	.50	.50
3	1.03	.99	.96	.93	.90	.88	.85	.83	.81	.79
4	1.38	1.35	1.31	1.28	1.25	1.22	1.19	1.16	1.14	1.12
5	1.73	1.70	1.66	1.63	1.60	1.57	1.54	1.51	1.48	1.46
6	2.08	2.05	2.02	1.98	1.95	1.92	1.89	1.86	1.83	1.80
7	2.43	2.40	2.36	2.33	2.30	2.27	2.24	2.21	2.18	2.15
8	2.79	2.76	2.72	2.69	2.66	2.62	2.59	2.56	2.53	2.50
9	3.15	3.11	3.08	3.04	3.01	2.97	2.94	2.91	2.88	2.85
10	3.51	3.47	3.44	3.40	3.37	3.33	3.30	3.27	3.23	3.20
11	3.86	3.82	3.79	3.75	3.72	3.68	3.65	3.62	3.58	3.55
12	4.21	4.18	4.15	4.11	4.08	4.04	4.01	3.97	3.94	3.90

RADII OF GYRATION FOR SQUARE COLUMNS.

OUTER DIAMETER ACROSS FLATS IN INCHES.	THICKNESS IN INCHES VARYING BY TENTHS.									
	.1	.2	.3	.4	.5	.6	.7	.8	.9	1.0
	CORRESPONDING RADIUS OF GYRATION IN INCHES.									
2	.78	.74	.71	.68	.65	.63	.61	.59	.58	.58
3	1.18	1.14	1.11	1.08	1.04	1.01	.98	.96	.93	.91
4	1.59	1.55	1.51	1.47	1.44	1.41	1.38	1.35	1.32	1.29
5	2.00	1.96	1.92	1.89	1.85	1.81	1.78	1.75	1.71	1.68
6	2.41	2.37	2.33	2.29	2.25	2.21	2.18	2.15	2.11	2.08
7	2.82	2.78	2.74	2.70	2.66	2.62	2.58	2.55	2.51	2.48
8	3.23	3.19	3.15	3.11	3.07	3.03	2.99	2.96	2.92	2.89
9	3.63	3.59	3.55	3.51	3.48	3.44	3.40	3.36	3.32	3.29
10	4.04	4.00	3.96	3.92	3.88	3.84	3.80	3.77	3.73	3.70
11	4.45	4.41	4.37	4.33	4.29	4.25	4.21	4.17	4.13	4.10
12	4.86	4.82	4.78	4.74	4.70	4.66	4.62	4.58	4.54	4.51

No. 17.

ROUND COLUMNS.

GREATEST SAFE LOADS IN LBS. PER SQ. IN. OF SECTION.

By this table for the same ratios of $\frac{l}{r}$ the safe loads are increased 10 per cent. over the results obtained for previous tables, as given in table No. 2.

Size Outer Diameter.	Condition of Ends.	2	4	6	8	10	12	14	16	18
12" Diameter. ¾" thick. R = 3·94	Fixed Ends....				15220	14330	13350	12640	12040	11470
	Flat Ends......				15220	14330	13350	12640	12040	11470
	Hinged Ends...				14680	13700	12670	12000	11250	10640
	Round Ends....				13940	12840	11660	10890	9950	9200
10" Diameter. ¾" thick. R = 3·37	Fixed Ends....			15660	14630	13490	12640	11940	11280	10640
	Flat Ends......			15660	14630	13490	12640	11940	11280	10640
	Hinged Ends...			15160	14030	12810	12000	11140	10450	9750
	Round Ends....			14470	13200	11820	10890	9820	8960	8170
8" Diameter. ¾" thick. R = 2·66	Fixed Ends...			14770	13490	12440	11570	10730	9940	9200
	Flat Ends......			14770	13490	12440	11570	10730	9940	9200
	Hinged Ends..			14190	12810	11680	10740	9850	8970	8170
	Round Ends....			13380	11820	10480	9330	8280	7330	6460
6" Diameter. ¾" thick. R = 2·00	Fixed Ends....		15220	13490	12140	11000	9940	9050	8440	7860
	Flat Ends....		15220	13490	12140	11000	9940	9040	8400	7650
	Hinged Ends..		14680	12810	11360	10150	8970	7960	7110	6310
	Round Ends....		13940	11820	10080	8600	7330	6220	5310	4490
5" Diameter. ¾" thick. R = 1·44	Fixed Ends....		14470	12540	11090	9850	8840	8150	7460	6740
	Flat Ends......		14470	12540	11090	9850	8820	8060	7260	6270
	Hinged Ends...		13870	11790	10250	8880	7660	6710	5920	4920
	Round Ends....		13020	10620	8720	7230	5790	4880	4130	3150
4" Diameter. ¾" thick. R = 1·33	Fixed Ends....		13490	11570	9940	8740	7860	7040	6190	5400
	Flat Ends......		13490	11570	9940	8710	7650	6560	5640	4680
	Hinged Ends...		12610	10740	8970	7520	6310	5240	4290	3260
	Round Ends....		11820	9330	7330	5750	4490	3460	2580	1880
3" Diameter. ⁹⁄₁₆" thick. R = 1·00	Fixed Ends....	15220	12140	9940	8440	7330	6190	5110	4130	3300
	Flat Ends......	15220	12140	9940	8400	6880	5640	4400	3440	2780
	Hinged Ends...	14680	11360	8970	7110	5560	4290	2990	2160	1610
	Round Ends....	13940	10080	7330	5310	3780	2580	1720	1230	910
2" Diameter. ¾" thick. R = ·66	Fixed Ends....	13490	9850	7820	6140	4510	3230	2290	1760	1340
	Flat Ends......	13490	9850	7590	5580	3770	2720	2020	1440	990
	Hinged Ends...	12810	8880	6240	4220	2420	1570	1100	790	600
	Round Ends....	11820	7230	4430	2540	1390	890	630	440	330

No. 17.

ROUND COLUMNS.

GREATEST SAFE LOADS IN LBS. PER SQ. IN. OF SECTION.

The calculations are based on the thicknesses and radii of gyration marked under the diameters on marginal columns. See description.

20	22	24	26	28	30	32	34	36	Condition of Ends.	Size Outer Diameter.
10910	10370	9850	9350	8990	8640	8340	8050	7770	Fixed Ends....	12"
10910	10370	9650	9350	8980	8610	8290	7930	7520	Flat Ends......	Diameter.
10050	9460	8880	8330	7880	7390	6970	6570	6180	Hinged Ends...	1" thick.
8490	7850	7230	6640	6030	5610	5150	4750	4370	Round Ends....	R = 3·94
10020	9430	8990	8620	8250	7910	7600	7280	6940	Fixed Ends....	10"
10020	9430	8980	8610	8190	7720	7260	6830	6460	Flat Ends......	Diameter.
9070	8430	7880	7390	6840	6380	5930	5510	5130	Hinged Ends....	1" thick,
7430	6740	6030	5610	5010	4560	4130	3730	3350	Round Ends....	R = 3·37
8740	8290	7860	7460	7040	6610	6190	5790	5400	Fixed Ends....	8"
8710	8250	7650	7070	6560	6110	5640	5140	4680	Flat Ends.....	Diameter.
7520	6900	6310	5920	5240	4780	4290	3750	3200	Hinged Ends...	1" thick.
5750	5090	4490	3960	3460	3000	2580	2210	1880	Round Ends....	R = 2·66
7330	6740	6190	5660	5110	4580	4130	3700	3300	Fixed Ends....	6"
6880	6270	5640	4990	4400	3850	3440	3080	2780	Flat Ends......	Diameter.
5560	4920	4290	3580	2990	2470	2160	1880	1610	Hinged Ends...	1" thick.
3780	3150	2580	2090	1720	1440	1230	1040	910	Round Ends....	R = 2·00
6100	5440	4760	4210	3670	3160	2790	2420	2110	Fixed Ends....	5"
5530	4730	4020	3500	3050	2680	2380	2110	1870	Flat Ends.....	Diameter.
4160	3310	2640	2220	1850	1540	1330	1150	1000	Hinged Ends...	1" thick.
2490	1900	1520	1260	1030	860	750	660	580	Round Ends....	R = 1·64
4580	3880	3260	2770	2320	2000	1780	1560	1360	Fixed Ends....	4"
3850	3210	2750	2370	2040	1740	1470	1220	1010	Flat Ends......	Diameter.
2470	2000	1590	1370	1110	940	800	690	610	Hinged Ends...	1" thick.
1440	1110	900	740	630	530	450	380	330	Round Ends....	R = 1·33
2650	2100	1790	1500	1240	1070	910	770	.....	Fixed Ends....	3"
2270	1850	1480	1150	910	710	530	440	.....	Flat Ends.. ...	Diameter.
1250	1000	810	670	560	460	350	280	.....	Hinged Ends...	³⁄₁₆" thick.
710	580	450	370	290	250	200	170	...	Round Ends...	R = 1·00
1040	810	.....	.....	.....	.....	.....	.....	.....	Fixed Ends....	2"
680	470	.....	.....	.....	.....	.....	.....	.....	Flat Ends......	Diameter.
440	300	.....	.....	.....	.....	.....	.....	.....	Hinged Ends...	1" thick.
240	180	.....	.....	.....	.....	.....	.....	.....	Round Ends...	R = ·66

No. 18.

SQUARE COLUMNS.

GREATEST SAFE LOAD IN LBS. PER SQUARE INCH OF SECTION.

By this table for the same ratios of $\dfrac{l}{r}$, the safe loads are increased 5 per cent. over the results obtained in table No. 2.

SIZE OF COLUMN.	CONDITION OF ENDS.	LENGTH IN FEET.								
		2	4	6	8	10	12	14	16	18
2″ ¼″ thick.. R = ·77	Fixed Ends....	13540	10330	8160	6760	5410	4130	3090	2310	1790
	Flat Ends......	13540	10330	8120	6320	4770	3440	2600	2020	1510
	Hinged Ends...	12910	9500	6920	5060	3420	2180	1500	1100	820
	Round Ends....	12100	8010	5210	3360	2000	1250	850	630	450
3″ 1 3/16″ thick R = 1·15	Fixed Ends....	14950	12160	10330	8690	7690	6760	5830	4920	4080
	Flat Ends......	14950	12160	10330	8680	7570	6320	5280	4240	3410
	Hinged Ends...	14480	11460	9500	7660	6290	5060	3980	2900	2160
	Round Ends....	13810	10400	8010	6020	4540	3360	2380	1680	1240
4″ ¼″ thick.. R — 1·53	Fixed Ends....	......	13540	11690	10330	9010	8150	7420	6720	6040
	Flat Ends......	......	13540	11690	10330	9010	8110	7180	6220	5540
	Hinged Ends...	......	12940	10940	9500	8050	6920	5900	5010	4260
	Round Ends...	......	12100	9750	8010	6440	5210	4180	3310	2590
5″ ¼″ thick.. R — 1·89	Fixed Ends....	......	14390	12610	11310	10250	9170	8400	7780	7260
	Flat Ends......	......	14390	12610	11310	10250	9170	8370	7700	6930
	Hinged Ends...	......	13860	11950	10540	9410	8220	7260	6400	5660
	Round Ends...	......	13120	10960	9260	7910	6630	5460	4660	3950
6″ ¼″ thick.. R — 2·30	Fixed Ends....	......	14950	13540	12160	11220	10330	9410	8690	8160
	Flat Ends......	......	14950	13540	12160	11220	10330	9410	8680	8120
	Hinged Ends...	......	14480	12940	11400	10450	9500	8480	7660	6920
	Round Ends....	......	13820	12100	10390	9150	8010	6910	6020	5210
8″ ¼″ thick.. R — 3·07	Fixed Ends....	......	......	14670	13540	12480	11690	10950	10250	9650
	Flat Ends......	......	......	14670	13540	12480	11690	10950	10250	9650
	Hinged Ends...	......	......	14170	12940	11800	10940	10160	9410	8750
	Round Ends...	......	......	13470	12100	10900	9750	8790	8010	7190
10″ ¼″ thick.. R — 3·87	Fixed Ends ...	......	......	......	14380	13540	12750	12060	11400	10860
	Flat Ends......	......	......	......	14380	13540	12750	12060	11400	10860
	Hinged Ends...	......	......	......	13960	12940	12090	11360	10640	10070
	Round Ends...	......	......	......	13120	12100	11130	10270	9380	8670
12″ ¼″ thick.. R — 4·66	Fixed Ends....	......	......	......	14950	14250	13420	12750	12140	11690
	Flat Ends......	......	......	......	14950	14250	13420	12750	12140	11690
	Hinged Ends...	......	......	......	14480	13700	12800	12090	11460	10940
	Round Ends....	......	......	......	13820	12950	11930	11130	10400	9750

No. 18.

SQUARE COLUMNS

GREATEST SAFE LOAD IN LBS. PER SQUARE INCH OF SECTION.

The calculations are based on the thicknesses and radii of gyration, marked under the diameters in marginal columns. See preceding description.

LENGTH IN FEET. 20	22	24	26	28	30	32	34	36	CONDITION OF ENDS.	SIZE OF COLUMN.
1440	1120	930	760	.....	.....	.....	.....	.....	Fixed Ends....	2"
1100	810	580	430	.....	.....	.....	.....	.....	Flat Ends.....	¼" thick.
640	490	380	270	.....	.....	.....	.....	.....	Hinged Ends...	R — ·77
360	260	210	170	.....	.....	.....	.....	.....	Round Ends....	
3380	2770	2290	1910	1660	1430	1210	1060	910	Fixed Ends....	3"
2820	2360	2010	1670	1370	1090	880	710	560	Flat Ends......	⁵⁄₁₆" thick
1690	1320	1090	900	750	630	550	450	370	Hinged Ends...	R — 1·15
950	760	630	510	420	360	290	240	210	Round Ends....	
5370	4670	4080	3540	3020	2650	2250	1960	1770	Fixed Ends....	4"
4710	3980	3110	2940	2560	2270	1980	1730	1500	Flat Ends	⅜" thick.
3370	2650	2160	1800	1470	1260	1080	930	810	Hinged Ends...	R — 1·52
1950	1530	1240	1000	830	710	620	540	450	Round Ends....	
6670	6130	5580	5020	4500	4020	3570	3150	2790	Fixed Ends....	5"
6230	5650	4970	4340	3800	3350	2970	2660	2370	Flat Ends......	⅜" thick.
4960	4370	3630	2990	2480	2120	1830	1540	1330	Hinged Ends...	R — 1·59
3460	2680	2140	1720	1440	1210	1010	870	760	Round Ends....	
7690	7210	6760	6310	5830	5410	4920	4500	4080	Fixed Ends....	6"
7570	6880	6320	5840		4770	4240	3800	3410	Flat Ends......	⅜" thick.
6280	5610	5060	4570	3980	3420	2900	2480	2160	Hinged Ends...	R — 2·30
4540	3900	3360	2870	2380	2000	2390	1440	1240	Round Ends....	
9010	8550	8160	7820	7470	7130	6760	6390	6040	Fixed Ends....	8"
9010	8530	8120	7760	7250	6750	6320	5930	5540	Flat Ends......	⅜" thick.
8050	7450	6920	6470	5960	5480	5060	4660	4260	Hinged Ends...	R — 3·07
6430	5690	5210	4730	4230	3780	3360	2960	2590	Round Ends....	
10330	9820	9320	8790	8490	8200	7920	7640	7340	Fixed Ends....	10"
10330	9820	9320	8790	8470	8170	7870	7500	7060	Flat Ends......	½" thick.
9500	8940	8400	7800	7380	6980	6590	6220	5780	Hinged Ends...	R — 3·87
8010	7390	6810	6170	5620	5230	4860	4480	4060	Round Ends....	
11130	10680	10250	9730	9320	8920	8640	8350	8110	Fixed Ends....	12"
11130	10680	10250	9730	9320	8920	8630	8320	8060	Flat Ends......	½" thick.
10350	9880	9410	8840	8400	7960	7600	7180	6860	Hinged Ends...	R — 4·55
9030	8440	7910	7300	6810	6340	5940	5490	5130	Round Ends....	

RIVETS AND PINS.

Rivets must be proportioned with sufficient bearing surface to resist crushing, and sufficient sectional area to resist shearing. Pins must be proportioned likewise, and also to safely resist the bending action which usually exists, owing to the centres of pressure being some distance from the centres of supports.

The effective bearing area of a rivet or pin is equal to its diameter multiplied by the thickness of the surface it bears on.

The shearing area is the area of the cross section of the pin or rivet for single shear, or double that section for double shear. For pins, the pressure on the pins multiplied by the leverage with which it acts on the pin supports is the bending moment. (See bending moments, page 78.)

The ultimate crushing strength of wrought iron is taken as equal to its tensile strength, viz., 50,000 lbs. per square inch, the shearing strength at $\frac{4}{5}$ of same, viz., 40,000 lbs. per square inch. The ultimate modulus of rupture is taken at 50,000, which is a fair estimate for cylindrical sections, as the average of many experiments we have made on that shape gives nearly that amount. The annexed table gives the ultimate resistance for single shear, or the area of the pin multiplied by 40,000, and the ultimate resistance to crushing, for each inch in thickness of bearing surface, or the diameter of the pin multiplied by 50,000.

The ultimate bending moments in inch lbs. correspond to the given diameter of pins, and are derived from the formula

$$M = \frac{50,000\,I}{\text{radius}},$$

which can be reduced to this form,

$$M = 6250 \times \text{area} \times \text{diameter, all in inches.}$$

To obtain the working resistances, these ultimate values must be divided by the factor of safety desirable to use.

The following proportions of the ultimate strength are commonly used for the purposes named.

For R. R. bridges,	$\frac{1}{6}$ of ultimate strength.
For light highway bridges,	$\frac{1}{4}$ of " "
For roof trusses, etc.,	$\frac{1}{3}$ of " "

Example.—A pin has its supports located three inches apart, and bears a load of 100,000 lbs. in the middle. What should the diameter of the pin be for a safety factor of five ?

$$\text{Bending moment} = \frac{100,000 \text{ lbs.} \times 3''}{4} = 75,000 \text{ inch lbs.}$$

The nearest diameter corresponding to this and taking $\frac{1}{5}$ of the tabular moments, is $4\frac{1}{4}$ inches.

The bearing value of this pin is ($\frac{1}{5}$ of table) 42,500 lbs. per inch of length, consequently the thickness of the metal which forms the pin bearings should be $\frac{100,000}{42500}$, or not less than 2.3 inches. For shear the pin has a large excess of strength, which will usually be found the case if properly proportioned otherwise.

11

ULTIMATE STRENGTH OF RIVETS AND PINS OF WROUGHT IRON.

For the working strength divide the tabular figures by the desired factor of safety.

Diameter in Inches of Rivet or Pin.	Area in Square Inches.	Ultimate Strength for Single Shear in Lbs.	Ultimate Crushing Strength per Inch Thickness of Bearing Surface.	Ultimate Bending Moment in Inch Lbs.
½	.196	7840	25000	614
9/16	.248	9920	28125	873
⅝	.307	12280	31250	1199
11/16	.371	14840	34375	1595
¾	.442	17680	37500	2073
13/16	.518	20720	40625	2632
⅞	.601	24040	43750	3287
1 inch.	.785	31400	50000	4906
⅛	.994	39760	56250	6993
¼	1.227	49080	62500	9586
⅜	1.485	59400	68750	12762
½	1.767	70680	75000	16566
⅝	2.074	82960	81250	21065
¾	2.405	96200	87500	26305
⅞	2.761	110440	93750	32357
2 inches.	3.141	125660	100000	39263
⅛	3.547	141880	106250	47109
¼	3.976	159040	112500	55913
⅜	4.430	177200	118750	65757
½	4.908	196320	125000	76688
⅝	5.412	216480	131250	88792
¾	5.940	237600	137500	102094
⅞	6.492	259680	143750	116825
3 inches.	7.068	282720	150000	132426
⅛	7.670	306800	156250	149694
¼	8.296	331840	162500	168514
⅜	8.946	357840	168750	188705
½	9.621	384840	175000	210459
⅝	10.321	412840	181250	233835
¾	11.045	441800	187500	258909
⅞	11.793	471720	193750	285618
4 inches.	12.566	502640	200000	314150
⅛	13.364	534560	206250	344540
¼	14.186	567440	212500	376816
⅜	15.033	601320	218750	411057
½	15.904	636160	225000	447300
⅝	16.800	672000	231250	485623
¾	17.721	708840	237500	526092
⅞	18.665	746600	243750	568700
5 inches.	19.635	785400	250000	613600
⅛	20.629	825160	256250	660773
¼	21.648	865920	262500	710326
⅜	22.691	907640	268750	762266
½	23.758	950320	275000	816667
⅝	24.850	994000	281250	873627
¾	25.967	1038680	287500	933189
⅞	27.109	1084360	293750	995410
6 inches.	28.274	1130960	300000	1060277

STRESSES IN SOME SIMPLE FORMS OF FRAMED STRUCTURES.

Compression indicated by the sign — and by solid lines. Tension by the sign + and by dotted lines.

When the prefix "stress" is used, the load borne by the member is indicated; otherwise the length of the member is meant.

CRANES.

Supported at the points A and B, maximum longitudinal stresses, due to weight W, suspended at the end. These stresses are modified by the position of the hoisting chain.

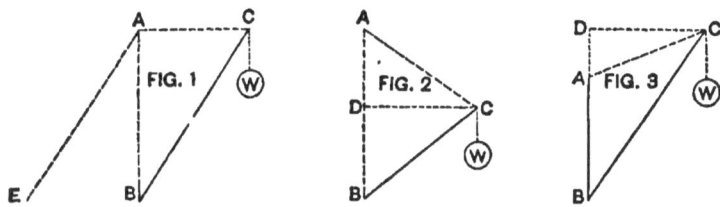

D is the point where a line drawn from C at right angles to $A B$ will intersect the latter.

$$\text{Stress } A\,C = + \frac{A\,C}{A\,B} \times W \qquad \text{Stress } B\,C = \frac{B\,C}{A\,B} \times W$$

$$" \quad A\,B = + \frac{A\,D}{A\,B} \times W \text{ in Fig. 2, or } = - \frac{A\,D}{A\,B} \times W \text{ in Fig. 3.}$$

When point A is supported by inclined back stays as shown in Fig. 1, and when the back stay is in the plane of $A\,B$ and W

$$\text{Stress } A\,E = + \frac{A\,C}{A\,B} \times W \times \frac{A\,E}{E\,B},$$

and a resulting compression ensues on

$$A\,B = - \frac{A\,C}{A\,B} \times W \times \frac{A\,B}{B\,E},$$

CRANES.

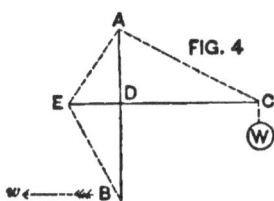

FIG. 4

Stress $C D = -\dfrac{C D}{A D} \times W$

"　$A C = +\dfrac{A C}{A D} \times W$

"　$E D = -$ stress $D C.$

Let $w =$ the horizontal reaction at B

$$w = \frac{C D}{A B} \times W$$

Stress $B E = +\dfrac{B E}{E D} \times w$

"　$A E = +\dfrac{A E}{D E} \times (\text{stress } C D - w)$

"　$B A = -\left(\dfrac{B D}{D E} \times w\right) + W$

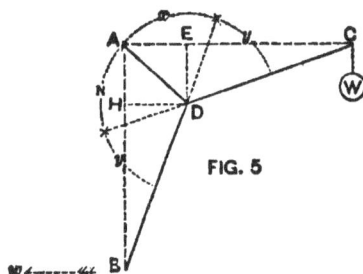

FIG. 5

E and H are points where lines drawn from D intersect at right angles $A C$ and $A B$. X, Y and Z are the angles formed by extending the braces $C D$ and $B D$ as indicated by dotted lines. $w =$ the horizontal reaction at B

$$w = \frac{A C}{A B} \times W.$$

Stress $A C = +\dfrac{C E}{E D} \times W.$　Stress $C D = -\dfrac{C D}{E D} \times W$

"　$A B = +\dfrac{B H}{D H} \times w.$　"　$B D = -\dfrac{B D}{H D} \times w$

"　$A D = -$ stress $C D \times \dfrac{\text{Sine } Y}{\text{Sine } X}$

or $= -$ stress $B D \times \dfrac{\text{Sine } Y}{\text{Sine } Z}$

Trussed Girders.

Weight in Middle.

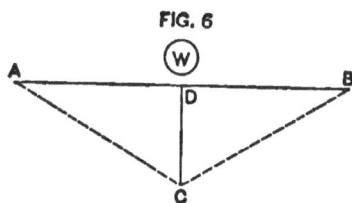

FIG. 6

Stress $A\,C$ or

$$B\,C = + \frac{A\,C}{D\,C} \times \frac{W}{2}$$

`` $A\,B = - \frac{A\,D}{D\,C} \times \frac{W}{2}$

`` $D\,C = - W$

———•●•———

Weight out of Centre.

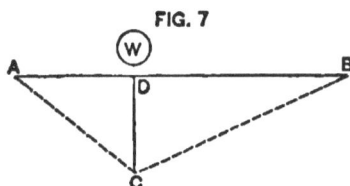

FIG. 7

Stress $A\,C = + \frac{A\,C \times D\,B}{A\,B \times D\,C} \times W$

`` $B\,C = + \frac{B\,C \times A\,D}{A\,B \times D\,C} \times W$

Stress $A\,B = - \frac{A\,D \times D\,B}{A\,B \times D\,C} \times W$

`` $D\,C = - W$

———•●•———

Equal Loads W. W.

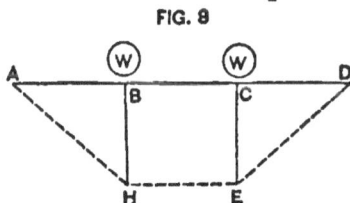

FIG. 8

Stress $A\,H$ or $D\,E = + \frac{A\,H}{B\,H} \times W$

`` $H\,E = + \frac{A\,B}{B\,H} \times W$

Stress $A\,D = - \frac{A\,B}{B\,H} \times W$

`` $B\,H$ or $C\,E = - W$

TRUSSED GIRDERS.

Unequal Loads W and w.

FIG. 9

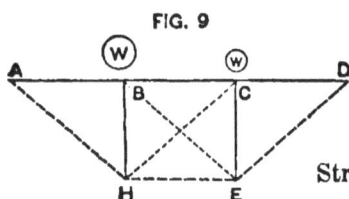

Stress as below on counter diagonals $B\,E$ or $H\,C$ according to position of greatest load.

$$\text{Stress } C\,H = +\,\frac{C\,H}{B\,H} \times \left(\frac{W-w}{3}\right)$$

• • •

Fink Truss.

FIG. 10

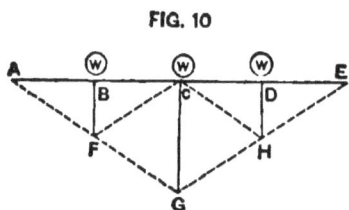

Stress $B\,F$ or $D\,H = -\,W$

Stress' $\quad C\,G = -\,2\,W$

" $\quad A\,E = -\,1\tfrac{1}{2}\,W \times \dfrac{A\,C}{C\,G}$

Stress $A\,F$ or $H\,E = +\,1\tfrac{1}{2}\,W \times \dfrac{A\,F}{F\,B}$

" $F\,G$ or $H\,G = +\,W \times \dfrac{A\,G}{C\,G}$

" $F\,C$ or $C\,H = +\,\dfrac{W}{2} \times \dfrac{A\,G}{C\,G}$

Roofs.

$w =$ load concentrated on each triangular apex.

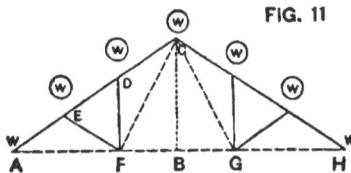

FIG. 11

Strut Stresses.

Stress $D F = - w$

" $E F = - \dfrac{w}{2} \times \dfrac{C H}{C B}$

Stresses on Ties.

Rafter Stresses.

Stress $F G = + 1\frac{1}{2} w \times \dfrac{B H}{B C}$ Stress $C E = - 2\ w \times \dfrac{C H}{C B}$

" $A F = + 2\frac{1}{2} w \times \dfrac{B H}{B C}$ " $E A = - 2\frac{1}{2} w \times \dfrac{C H}{C B}$

" $C F = + 1\frac{1}{2} w \times \dfrac{C G}{B C}$

Roofs.

$w =$ load concentrated on each triangular apex.

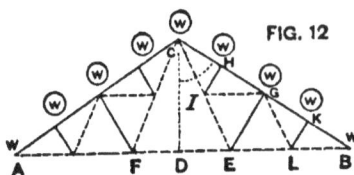

FIG. 12

Strut Stresses.

Stress $H I$ or $K L = - w \times \dfrac{D B}{C B}$

"　　　$G E = - 2 w \times \dfrac{D B}{C B}$

Rafter Stresses.

Stress $K B = - \left(\dfrac{7 w}{2} \times \dfrac{C B}{C D} \right)$

"　$G K = - \left(\dfrac{7 w}{2} \times \dfrac{C B}{C D} - w \times \dfrac{C D}{C B} \right)$

"　$H G = - \left(\dfrac{7 w}{2} \times \dfrac{C B}{C D} - 2 w \times \dfrac{C D}{C B} \right)$

"　$C H = - \left(\dfrac{7 w}{2} \times \dfrac{C B}{C D} - 3 w \times \dfrac{C D}{C B} \right)$

Stresses on Ties.

Stress $G I$ or $G L = + \dfrac{w}{2} \times \dfrac{D B}{C B} \times \dfrac{C B}{C D}$

"　　　　$E I = + w \times \dfrac{D B}{C B} \times \dfrac{C B}{C D}$

$C I = \times \dfrac{3 w}{2} \times \dfrac{D B}{C B} \times \dfrac{C B}{C D}$

$F E = + 8 w \times \dfrac{D B}{D C}$

$E L =$ the sum of the stresses on $F E$ and $E I$.

$L B =$ the sum of the stresses on $E L$ and $G L$.

Roofs.

$w =$ load concentrated on each triangular apex.

The rafters and horizontal tie being each uniformly subdivided.

Strut Stresses.

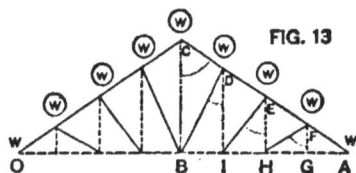

FIG. 13

Stress $FH = -\dfrac{w}{2} \times \dfrac{FH}{FG}$

" $EI = -\ w\ \times \dfrac{EI}{EH}$

" $DB = -\dfrac{3w}{2} \times \dfrac{DB}{DI}$

Vertical Ties.

Stress $EH = +\dfrac{w}{2}.$ Stress $DI = +w.$ Stress $CB = +3w.$

Rafter Stresses.

Stress $CD = -2\ w \times \dfrac{CA}{CB}$

" $DE = -2\tfrac{1}{4}w \times \dfrac{CA}{CB}$

" $EF = -3\ w \times \dfrac{CA}{CB}$

" $FA = -3\tfrac{1}{4}w \times \dfrac{CA}{CB}$

Horizontal Tie.

Stress at $B = +2\ w \times \dfrac{BA}{BC}$

" $BI = +$ stress at $B + \left(\text{stress } DB \times \dfrac{BI}{BD}\right)$

' $IH = +$ " $BI + \left(\ \text{"}\ EI \times \dfrac{HI}{EI}\right)$

" $HA = +$ " $IH + \left(\ \text{"}\ FH \times \dfrac{HG}{HF}\right)$

WROUGHT IRON SHAFTING.

(For steel shafting see page 29.)

The ultimate resistance of wrought iron to shearing averages about $\frac{8}{10}$ of its ultimate tensile strength, *i.e.*, about 40,000 lbs. per sq. inch of section. The torsional resistance of any wrought-iron shaft can be determined when the shearing resistance is known ; thus,

$$T = .196 \, d^3 s \text{ for round shafts,} \qquad (a)$$

$$T = .28 \, d^3 s \text{ for square shafts.} \qquad (b)$$

$d =$ diameter of the shaft in inches.

$s =$ shearing strength in lbs. per sq. inch.

$T =$ the torsional moment in inch-lbs., that is, the force in lbs. multiplied by the length in inches, of the lever through which the force acts.

Taking s at 40,000 lbs., and assuming that in machinery the working value of wrought iron should be taken at from one-fourth to one-fifth of its ultimate strength, these being factors of safety sanctioned by good practice, we adopt the mean of the two, which makes the working resistance to shearing $= 9,000$ lbs. per sq. inch. Putting this in terms of the torsional moment and of the diameter, we derive from equations a and b,

$$T = 1760 \, d^3 \text{ for round shafts,} \qquad (c)$$

$$T = 2520 \, d^3 \text{ for square shafts,} \qquad (d)$$

$$d = \sqrt[3]{\frac{T}{1760}} \text{ for round shafts,} \qquad (e)$$

$$d = \sqrt[3]{\frac{T}{2520}} \text{ for square shafts.} \qquad (f)$$

Example 1.—What should be the diameter of a round wrought

iron shaft to safely resist a force of 1,000 lbs. acting through a lever 30 inches long ?

(e) $$d = \sqrt[3]{\frac{1000 \times 30}{1760}} = 2.6 \text{ inches diameter.}$$

These formulæ apply to shafts subject to twisting strains alone. In practice, however, sucn cases seldom occur, as shafts are generally subjected to combined bending and twisting strains. As there are no experimental data for such a combination of forces, we have to rely on analysis, which gives the following:

$$T^1 = M + \sqrt{M^2 + T^2} \qquad (g)$$

$M =$ bending moments in inch-lbs. (See page 78.)
$T =$ twisting "   "
$T^1 =$ a *new* twisting moment which, substituted for T in equations (e) and (f), will give the desired proportions for the shaft.

In revolving shafts the longitudinal stress resulting from the bending action is continually changing from tension to compression, and vice versa.

It is therefore advisable, for reasons given on page 34, to increase the factor of safety as the bending stress increases comparatively to the torsional stress.

The following changes in factors of safety are recommended :

RATIO OF M TO T.	FACTOR OF SAFETY.	DIVISOR IN FORMULA (e).
$M = .3T$ or less,	$4\frac{1}{2}$	1760
$M = .6T$ "	5	1570
$M = T$ "	$5\frac{1}{2}$	1430
$M =$ greater than T,	6	1310

Example 2.—What should be the diameter of the journals of a wrought-iron shaft of a steam engine. The piston being 12 inches diam., crank 12 inches long, and the leverage from centre of crank to journal in the direction of the shaft being 6 inches, steam pressure 80 lbs. per sq. inch, making pressure on crank $= 9050$ lbs.?

$$T = 9050 \times 12 = 108600 \text{ inch-lbs.}$$

$$M = 9050 \times 6 = 54300 \quad ``$$

$(g) \quad T^1 = 54300 + \sqrt{54300^2 + 108600^2} = 175720 \text{ inch-lbs.}$

Substituting the above in equation (*e*), with the factor of safety as explained above,

$$d = \sqrt[3]{\frac{175720}{1570}} = 4.82 \text{ inches diameter.}$$

The following illustrates a case where the bending moment is greater than the twisting moment :

Example 3.—A non-continuous shaft is so located that it must have its bearings 84 inches apart, and carry in the middle a 60-inch pulley driven by a 12-inch belt, the effective weight at centre of shaft $= 600$ lbs., and the belt exercises a vertical pull of 1000 lbs. What is the proper diameter of the shaft ?

$$M = \frac{(1000 + 600) \times 84}{4} = 33600 \text{ inch-lbs. (see page 78).}$$

$$T = 1000 \times 30 = 30000 \text{ inch-lbs.}$$

$(g) \quad T^1 = 33600 + \sqrt{33600^2 + 30000^2} = 78640 \text{ inch-lbs.}$

As M is greater than T, use a factor of safety of 6, which becomes by equation (*e*),

$$d = \sqrt[3]{\frac{78640}{1310}} = 4.12 \text{ inches diam.}$$

If above shaft was continuous and uniformly loaded, the

bending moment would be less. (See Table of Bending Moments, page 80.)

HORSE POWER.

If it is desired to find the relations between horse power and diameters of shafts, the elements of time and velocity have to be considered. Taking the horse power HP at 396000 inch-lbs. per minute, we have $HP = \dfrac{6.28 \times T \times V}{396000}$, where $V =$ revolutions per minute.

(h) $$T = \frac{63057\,HP}{V},$$

or in terms of the diameter by equation (c) we get,

$$d = \sqrt[3]{\frac{36\,HP}{V}} \cdot \qquad (i)$$

The above will give the proper diameter of a shaft for transmitting any desired HP when the shaft is subjected to twisting stress alone, but, as previously stated, such a case seldom occurs, we must combine the bending and twisting stresses, for which a general rule will be given at the close of the subject.

DEFLECTION OF SHAFTING.

For continuous line shafting used for transmitting power in shops, factories, etc., it is considered good practice to limit the deflection to a maximum of $\frac{1}{100}$ of an inch per foot of length. The weight of bare shafting in lbs. $= 2.6\,d^2l = W$, or when as fully loaded with pulleys as is customary in practice, and allowing 40 lbs. per inch of width for the vertical pull of the belts, experience shows the load in lbs. to be about $13\,d^2l = W$. Taking the modulus of transverse elasticity at 26,000,000 lbs., we can derive from the authoritative formulæ the following :

$$l = \sqrt[3]{873d^2} \text{ for bare shafts,} \qquad (j)$$

$$l = \sqrt[3]{175d^2} \text{ for shafts carrying pulleys, etc.,} \qquad (k)$$

which would be the maximum distance in feet between bearings for continuous shafting subjected to bending stress alone.

If the length is fixed, and we desire the diameter of the shaft, we have,

$$d = \sqrt{\frac{l^2}{873}} \text{ for bare shafting,} \qquad (l)$$

$$d = \sqrt{\frac{l^2}{175}} \text{ for shafting carrying pulleys, etc.} \qquad (m)$$

To apply the above to revolving shafting subjected to both twisting and bending stress, it is necessary to combine equations (j) and (k) with equation (i).

But in shafting, with the same transmission of power, the torsional stress is inversely proportional to the velocity of rotation, while the bending stress will not be reduced in the same ratio. It is, therefore, impossible to write a formula covering the whole problem and sufficiently simple for practical application, but the following rules are correct within the range of velocities usual in practice.

WORKING FORMULÆ FOR CONTINUOUS SHAFTING.

For the diameter (d) in inches, and the maximum length (l) in feet between bearings of wrought-iron shafting so proportioned as to deflect not more than $\frac{1}{100}$ of an inch per foot of length, allowance being made for the weakening effect of key seats,

$$d = \sqrt[3]{\frac{50\,HP}{V}} \text{ for bare shafts,} \qquad (n)$$

$$d = \sqrt[3]{\frac{70\,HP}{V}} \text{ for shafts carrying pulleys, etc.,} \qquad (o)$$

$$l = \sqrt[3]{720\,d^2} \text{ for bare shafts,} \qquad (p)$$

$$l = \sqrt[3]{140\,d^2} \text{ for shafts carrying pulleys, etc.,} \qquad \cdot(q)$$

In the event of the whole power being received on a principal shaft, the proper size of the shaft can be estimated direct by formula (g).

Example 4.—A principal shaft receiving 150 HP from the engine, revolves 150 R. P. M., and is continuous over bearings located 6 feet apart, the centre of main pulley being 24 inches from one bearing and 48 inches from the other. The effective loa ᐟ at the centre of the pulley resulting from weight of pulley and shaft, and tension of belt, is 1500 lbs. What should be the diameter of the shaft ?

Note.—Excepting special cases which rarely occur in practice, it is best to treat such shafts as non-continuous.

By rule 5, page 79, we have,

$$M = \frac{1500 \times 24 \times 48}{72} = 24000 \text{ inch-lbs,}$$

and by formula (h) we have,

$$T = \frac{63000 \times 150}{150} = 63000 \text{ inch-lbs.,}$$

then, by formula (g) we have

$$T' = 24000 + \sqrt{24000^2 + 63000^2} = 92290 \text{ inch-lbs.}$$

and by formula (e),

$$d = \sqrt[3]{\frac{92290}{1760}} = 3.74 \text{ inches.}$$

BELTING.

When designing shafting, allow for the tension of belting, 50 lbs. per inch of width for single leather belt or its equivalent, or 80 lbs. per inch of width for double leather belt, or its equivalent of other material.

WORKING PROPORTIONS FOR CONTINUOUS SHAFTING.

TRANSMITTING POWER, BUT SUBJECT TO NO BENDING ACTION EXCEPT ITS OWN WEIGHT.

Diameter of Shaft in Inches.	Max. Safe Torsional Moment in Inch-Pounds.	Revolutions per Minute. 100	150	200	Max. Distance in Feet between Bearings.
		HP	*HP*	*HP*	
$1\frac{1}{2}$	5940	6	10	14	11.7
$1\frac{5}{8}$	7552	9	13	17	12.4
$1\frac{3}{4}$	9432	11	16	21	13.0
$1\frac{7}{8}$	11602	13	20	26	13.6
2	14080	16	24	32	14.2
$2\frac{1}{8}$	16892	19	29	38	14.8
$2\frac{1}{4}$	20048	23	34	46	15.4
$2\frac{3}{8}$	23580	27	40	54	16.0
$2\frac{1}{2}$	27500	31	47	63	16.5
$2\frac{3}{4}$	36603	42	62	83	17.6
3	47520	54	81	108	18.6
$3\frac{1}{4}$	60417	69	103	137	19.7
$3\frac{1}{2}$	75460	86	129	172	20.7
$3\frac{3}{4}$	92812	105	158	211	21.6
4	112640	128	192	256	22.6

WORKING PROPORTIONS FOR CONTINUOUS SHAFTING.

TRANSMITTING POWER, AND SUBJECT TO BENDING ACTION OF PULLEYS, BELTING, ETC.

DIAMETER OF SHAFT IN INCHES.	MAX. SAFE TORSIONAL MOMENT IN INCH-POUNDS.	REVOLUTIONS PER MINUTE.			MAX. DISTANCE IN FEET BETWEEN BEARINGS.
		100	150	200	
		HP	HP	HP	
$1\frac{1}{2}$	5940	5	7	10	6.8
$1\frac{5}{8}$	7552	6	9	12	7.2
$1\frac{3}{4}$	9432	8	11	15	7.5
$1\frac{7}{8}$	11602	9	14	19	7.9
2	14080	11	17	23	8.2
$2\frac{1}{8}$	16892	14	21	27	8.6
$2\frac{1}{4}$	20048	16	24	33	8.9
$2\frac{3}{8}$	23580	19	29	38	9.2
$2\frac{1}{2}$	27500	22	33	45	9 6
$2\frac{3}{4}$	36603	24	36	48	10.2
3	47520	39	58	77	10.8
$3\frac{1}{4}$	60417	49	71	98	11.4
$3\frac{1}{2}$	75460	61	92	123	12.0
$3\frac{3}{4}$	92812	75	113	151	12.5
4	112640	91	137	183	13.1

TABLE OF CIRCLES.

Circumferences or areas intermediate of those in the table, may be found by simple arithmetical proportion. The diameters, etc., are in inches; but it is plain that if the diameters are taken as feet, yards, etc., the other parts will also be in those same measures.

DIAM. INS.	CIRCUMF. INS.	AREA. SQ. INS.	DIAM. INS.	CIRCUMF. INS.	AREA. SQ. INS.	DIAM. INS.	CIRCUMF. INS.	AREA. SQ. INS.
1-64	.049087	.00019	1 15-16	6.08684	2.9483	4 15-16	15.5116	19.147
1-32	.098175	.0 077	2.	6.28319	3.1416	5.	15.7080	19.635
3-64	.147262	.00173	1-16	6.47953	3.3410	1-16	15.9043	20.129
1-16	.196350	.00307	1-8	6.67588	3.5466	1-8	16.1007	20.629
3-32	.294524	.00690	3-16	6.87223	3.7583	3-16	16.2970	21.135
1-8	.392699	.01227	1-4	7.06858	3.9761	1-4	16.4934	21.648
5-32	.490874	.01917	5-16	7.26493	4.2000	5-16	16.6897	22.166
3-16	.589049	.02761	3-8	7.46128	4.4301	3-8	16.8861	22.691
7-32	.687223	.03758	7-16	7.65763	4.6664	7-16	17.0824	23.221
1-4	.785398	.04909	1-2	7.85398	4.9087	1-2	17.2788	23.758
9-32	.883573	.06213	9-16	8.05033	5.1572	9-16	17.4731	24.301
5-16	.981748	.07670	5-8	8.24668	5.4119	5-8	17.6715	24.850
11-32	1.07992	.09281	11-16	8.44303	5.6727	11-16	17.8678	25.406
3-8	1.17810	.11045	3-4	8.63938	5.9396	3-4	18.0642	25.967
13-32	1.27627	.12962	13-16	8.83573	6.2126	13-16	18.2605	26.535
7-16	1.37445	.15033	7-8	9.032 8	6.4918	7-8	18.4569	27.109
15-32	1.47262	.17257	15-16	9.22243	6.7771	15-16	18.6532	27.688
1-2	1.57080	.19635	3.	9.42478	7.0686	6.	18.8496	28.274
17-32	1.66897	.22166	1-16	9.62113	7.3662	1-8	19.2423	29.465
9-16	1.76715	.24850	1-8	9.81750	7.6699	1-4	19.6350	30.680
19-32	1.86532	.27688	3-16	10.0138	7.9798	3-8	20.0277	31.919
5-8	1.96350	.30680	1-4	10.2102	8.2958	1-2	20.4204	33.183
21-32	2.06167	.33824	5-16	10.4065	8.6179	5-8	20.8131	34.472
11-16	2.15984	.37122	3-8	10.6029	8.9462	3-4	21.2058	35.785
23-32	2.25802	.40574	7-16	10.7992	9.2806	7-8	21.5984	37.122
3-4	2.35619	.44179	1-2	10.9956	9.6211	7.	21.9911	38.485
25-32	2.45437	.47937	9-16	11.1919	9.9678	1-8	22.3838	39.871
13-16	2.55254	.51849	5-8	11.3883	10.321	1-4	22.7765	41.282
27-32	2.65072	.55914	11-16	11.5846	10.680	3-8	23.1692	42.718
7-8	2.74889	.60132	3-4	11.7810	11.045	1-2	23.5619	44.179
29-32	2.84707	.64504	13-16	11.9773	11.416	5-8	23.9546	45.664
15-16	2.94524	.69029	7-8	12.1737	11.793	3-4	24.3473	47.173
31-32	3.04342	.73708	15-16	12.3700	12.177	7-8	24.7400	48.707
1.	3.14159	.78540	4.	12.5664	12.566	8.	25.1327	50.265
1-16	3.33794	.88664	1-16	12.7627	12.962	1-8	25.5254	51.849
1-8	3.53429	.99402	1-8	12.9591	13.364	1-4	25.9181	53.456
3-16	3.73064	1.1075	3-16	13.1554	13.772	3-8	26.3108	55.088
1-4	3.92699	1.2272	1-4	13.3518	14.186	1-2	26.7035	56.745
5-16	4.12334	1.3530	5-16	13.5481	14.607	5-8	27.0962	58.426
3-8	4.31969	1.4849	3-8	13.7445	15.033	3-4	27.4889	60.132
7-16	4.51604	1.6230	7-16	13.9108	15.466	7-8	27.8816	61.862
1-2	4.71239	1.7671	1-2	14.1372	15.904	9.	28.2743	63.617
9-16	4.90874	1.9175	9-16	14.3335	16.349	1-8	28.6670	65.397
5-8	5.10509	2.0739	5-8	14.5299	16.800	1-4	29.0597	67.201
11-16	5.30144	2.2365	11-16	14.7262	17.257	3-8	29.4524	69.029
3-4	5.49779	2.4053	3-4	14.9226	17.721	1-2	29.8451	70.882
13-16	5.69414	2.5802	13-16	15.1189	18.190	5-8	30.2378	72.760
7-8	5.89049	2.7612	7-8	15.3153	18.665	3-4	30.6305	74.662

TABLE OF CIRCLES—Continued.

DIAM. INS.	CIRCUMF. INS.	AREA. SQ. INS.	DIAM. INS.	CIRCUMF. INS.	AREA. SQ. INS.	DIAM. INS.	CIRCUMF. INS.	AREA. SQ. INS.
9 7-8	31.0232	76.589	16 3-4	52.6217	220.35	23 5-8	74.2201	438.36
10.	31.4159	78.540	7-8	53.0144	223.65	3-4	74.6128	443.01
1-8	31.8086	80.516	17.	53.4071	226.98	7-8	75.0055	447.69
1-4	32.2013	82.516	1-8	53.7998	230.33	24.	75.3982	452.39
3-8	32.5940	84.541	1-4	54.1925	233.71	1-8	75.7909	457.11
1-2	32.9867	86.590	3-8	54.5852	237.10	1-4	76.1836	461.86
5-8	33.3794	88.664	1-2	54.9779	240.53	3-8	76.5763	466.64
3-4	33.7721	90.763	5-8	55.3706	243.98	1-2	76.9690	471.44
7-8	34.1648	92.886	3-4	55.7633	247.45	5-8	77.3617	476.26
11.	34.5575	95.033	7-8	56.1560	250.95	3-4	77.7544	481.11
1-8	34.9502	97.205	18.	56.5487	254.47	7-8	78.1471	485.98
1-4	35.3429	99.402	1-8	56.9414	258.02	25.	78.5398	490.87
3-8	35.7356	101.62	1-4	57.3341	261.59	1-8	78.9325	495.79
1-2	36.1283	103.87	3-8	57.7268	265.18	1-4	79.3252	500.74
5-8	36.5210	106.14	1-2	58.1195	268.80	3-8	79.7179	505.71
3-4	36.9137	108.43	5-8	58.5122	272.45	1-2	80.1106	510.71
7-8	37.3064	110.75	3-4	58.9049	276.12	5-8	80.5033	515.72
12.	37.6991	113.10	7-8	59.2976	279.81	3-4	80.8960	520.77
1-8	38.0918	115.47	19.	59.6903	283.53	7-8	81.2887	525.84
1-4	38.4845	117.86	1-8	60.0830	287.27	26.	81.6814	530.93
3-8	38.8772	120.28	1-4	60.4757	291.04	1-8	82.0741	536.05
1-2	39.2699	122.72	3-8	60.8684	294.83	1-4	82.4668	541.19
5-8	39.6626	125.19	1-2	61.2611	298.65	3-8	82.8595	546.35
3-4	40.0553	127.68	5-8	61.6538	302.49	1-2	83.2522	551.55
7-8	40.4480	130.19	3-4	62.0465	306.35	5-8	83.6449	556.76
13.	40.8407	132.73	7-8	62.4392	310.24	3-4	84.0376	562.00
1-8	41.2334	135.30	20.	62.8319	314.16	7-8	84.4303	567.27
1-4	41.6261	137.89	1-8	63.2246	318.10	27.	84.8230	572.56
3-8	42.0188	140.50	1-4	63.6173	322.06	1-8	85.2157	577.87
1-2	42.4115	143.14	3-8	64.0100	326.05	1-4	85.6084	583.21
5-8	42.8042	145.80	1-2	64.4026	330.06	3-8	86.0011	588.57
3-4	43.1969	148.49	5-8	64.7953	334.10	1-2	86.3938	593.96
7-8	43.5896	151.20	3-4	65.1880	338.16	5-8	86.7865	599.37
14.	43.9823	153.94	7-8	65.5807	342.25	3-4	87.1792	604.81
1-8	44.3750	156.70	21.	65.9734	346.36	7-8	87.5719	610.27
1-4	44.7677	159.48	1-8	66.3661	350.50	28.	87.9646	615.75
3-8	45.1604	162.30	1-4	66.7588	354.66	1-8	88.3573	6 1.26
1-2	45.5531	165.13	3-8	67.1515	358.84	1-4	88.7500	626.80
5-8	45.9458	167.99	1-2	67.5442	363.05	3-8	89.1427	632.36
3-4	46.3385	170.87	5-8	67.9369	367.28	1-2	89.5354	637.94
7-8	46.7312	173.78	3-4	68.3296	371.54	5-8	89.9281	643.55
15.	47.1239	176.71	7-8	68.7223	375.83	3-4	90.3208	649.18
1-8	47.5166	179.67	22.	69.1150	380.13	7-8	90.7135	654.84
1-4	47.9093	182.65	1-8	69.5077	384.46	29.	91.1062	660.52
3-8	48.3020	185.66	1-4	69.9004	388.82	1-8	91.4989	666.23
1-2	48.6947	188.69	3-8	70.2931	393.20	1-4	91.8916	671.96
5-8	49.0874	191.75	1-2	70.6858	397.61	3-8	92.2843	677.71
3-4	49.4801	194.83	5-8	71.0785	402.04	1-2	92.6770	683.49
7-8	49.8728	197.93	3-4	71.4712	406.49	5-8	93.0697	689.30
16.	50.2655	201.06	7-8	71.8639	410.97	3-4	93.4624	695.13
1-8	50.6582	204.22	23.	72.2566	415.48	7-8	93.8551	700.98
1-4	51.0509	207.39	1-8	72.6493	420.00	30.	94.2478	706.86
3-8	51.4436	210.60	1-4	73.0420	424.56	1 8	94.6405	712.76
1-2	51.8363	213.82	3-8	73.4347	429.13	1-4	95.0332	718.69
5-8	52.2290	217.08	1-2	73.8274	433.74	3-8	95.4259	724.64

TABLE OF CIRCLES—Continued.

DIAM. INS.	CIR- CUMF. INS.	AREA. SQ. INS.	DIAM. INS.	CIR- CUMF. INS.	AREA. SQ. INS.	DIAM. INS.	CIR- CUMF. INS.	AREA. SQ. INS.
30 1-2	95.8186	730.62	37 3-8	117.417	1097.1	44 1-4	139.015	1537.9
5-8	96.2113	736.62	1-2	117.810	1104.5	3-8	139.408	1546.6
3-4	96.6040	742.64	5-8	118.202	1111.8	1-2	139.801	1555.3
7-8	96.9967	748.69	3-4	118.596	1119.2	5-8	140.194	1564.0
31.	97.3894	754.77	7-8	118.988	1126.7	3-4	140.586	1572.8
1-8	97.7821	760.87	38.	119.381	1134.1	7-8	140.979	1581.6
1-4	98.1748	766.99	1-8	119.773	1141.6	45.	141.372	1590.4
3-8	98.5675	773.14	1-4	120.166	1149.1	1-8	141.764	1599.3
1-2	98.9602	779.31	3-8	120.559	1156.6	1-4	142.157	1608.2
5-8	99.3529	785.51	1-2	120.951	1164.2	3-8	142.550	1617.0
3-4	99.7456	791.73	5-8	121.344	1171.7	1-2	142.942	1626.0
7-8	100.138	797.98	3-4	121.737	1179.3	5-8	143.335	1634.9
32.	100.531	804.25	7-8	122.129	1186.9	3-4	143.728	1643.9
1-8	100.924	810.54	39.	122.522	1194.6	7-8	144.121	1652.9
1-4	101.316	816.86	1-8	122.915	1202.3	46.	144.513	1661.9
3-8	101.709	823.21	1-4	123.308	1210.0	1-8	144.906	1670.9
1-2	102.102	829.58	3-8	123.700	1217.7	1-4	145.299	1680.0
5-8	102.494	835.97	1-2	124.093	1225.4	3-8	145.691	1689.1
3-4	102.887	842.39	5-8	124.486	1233.2	1-2	146.084	1698.2
7-8	103.280	848.83	3-4	124.878	1241.0	5-8	146.477	1707.4
33.	103.673	855.30	7-8	125.271	1248.8	3-4	146.869	1716.5
1-8	104.065	861.79	40.	125.664	1256.6	7-8	147.262	1725.7
1-4	104.458	868.31	1-8	126.056	1264.5	47.	147.655	1734.9
3-8	104.851	874.85	1-4	126.449	1272.4	1-8	148.048	1744.2
1-2	105.243	881.41	3-8	126.842	1280.3	1-4	148.440	1753.5
5-8	105.636	888.00	1-2	127.235	1288.2	3-8	148.833	1762.7
3-4	106.029	894.62	5-8	127.627	1296.2	1-2	149.226	1772.1
7-8	106.421	901.26	3-4	128.020	1304.2	5-8	149.618	1781.4
34.	106.814	907.92	7-8	128.413	1312.2	3-4	150.011	1790.8
1-8	107.207	914.61	41.	128.805	1320.3	7-8	150.404	1800.1
1-4	107.600	921.32	1-8	129.198	1328.3	48.	150.796	1809.6
3-8	107.992	928.06	1-4	129.591	1336.4	1-8	151.189	1819.0
1-2	108.385	934.82	3-8	129.983	1344.5	1-4	151.582	1828.5
5-8	108.778	941.61	1-2	130.376	1352.7	3-8	151.975	1837.9
3-4	109.170	948.42	5-8	130.769	1360.8	1-2	152.367	1847.5
7-8	109.563	955.25	3-4	131.161	1369.0	5-8	152.760	1857.0
35.	109.956	962.11	7-8	131.554	1377.2	3-4	153.153	1866.5
1-8	110.348	969.00	42.	131.947	1385.4	7-8	153.545	1876.1
1-4	110.741	975.91	1-8	132.340	1393.7	49.	153.938	1885.7
3-8	111.134	982.84	1-4	132.732	1402.0	1-8	154.331	1895.4
1-2	111.527	989.80	3-8	133.125	1410.3	1-4	154.723	1905.0
5-8	111.919	996.78	1-2	133.518	1418.6	3-8	155.116	1914.7
3-4	112.312	1003.8	5-8	133.910	1427.0	1-2	155.509	1924.4
7-8	112.705	1010.8	3-4	134.303	1435.4	5-8	155.902	1934.2
36.	113.097	1017.9	7-8	134.696	1443.8	3-4	156.294	1943.9
1-8	113.490	1025.0	43.	135.068	1452.2	7-8	156.687	1953.7
1-4	113.883	1032.1	1-8	135.481	1460.7	50.	157.080	1963.5
3-8	114.275	1039.2	1-4	135.874	1469.1	1-8	157.472	1973.3
1-2	114.668	1046.3	3-8	136.267	1477.6	1-4	157.865	1983.2
5-8	115.061	1053.5	1-2	136.659	1486.2	3-8	158.258	1993.1
3-4	115.454	1060.7	5-8	137.052	1494.7	1-2	158.650	2003.0
7-8	115.846	1068.0	3-4	137.445	1503.3	5-8	159.043	2012.9
37.	116.239	1075.2	7-8	137.837	1511.9	3-4	159.436	2022.8
1-8	116.632	1082.5	44.	138.230	1520.5	7-8	159.829	2032.8
1-4	117.024	1089.8	1-8	138.623	1529.2	51.	160.221	2042.8

TABLE OF CIRCLES—*Continued.*

DIAM. INS.	CIR-CUMF. INS.	AREA. SQ. INS.	DIAM. INS.	CIR-CUMF. INS.	AREA. SQ. INS.	DIAM. INS.	CIR-CUMF. INS.	AREA. SQ. INS.
51 1-8	160.614	2052.8	58.	182.212	2642.1	64 7-8	203.811	3305.6
1-4	161.007	2062.9	1-8	182.605	2653.5	65.	204.204	3318.3
3-8	161.399	2073.0	1-4	182.998	2664.9	1-8	204.596	3331.1
1-2	161.792	2083.1	3-8	183.390	2676.4	1-4	204.989	3343.9
5-8	162.185	2093.2	1-2	183.783	2687.8	3-8	205.382	3356.7
3-4	162.577	2103.3	5-8	184.176	2699.3	1-2	205.774	3369.6
7-8	162.970	2113.5	3-4	184.569	2710.9	5-8	206.167	3382.4
52.	163.363	2123.7	7-8	184.961	2722.4	3-4	206.560	3395.3
1-8	163.756	2133.9	59.	185.354	2734.0	7-8	2 6.952	3408.2
1-4	164.148	2144.2	1-8	185.747	2745.6	66.	207.345	3421.2
3-8	164.541	2154.5	1-4	186.139	2757.2	1-8	207.738	3434.2
1-2	164.934	2164.8	3-8	186.532	2768.8	1-4	208.131	3447.2
5-8	165.326	2175.1	1-2	186.925	2780.5	3-8	208.523	3460.2
3-4	165.719	2185.4	5-8	187.317	2792.2	1-2	208.916	3473.2
7-8	166.112	2195.8	3-4	187.710	2803.9	5-8	209.309	3486.3
53.	166.504	2206.2	7-8	188.103	2815.7	3-4	209.701	3499.4
1-8	166.897	2216.6	60.	188.496	2827.4	7-8	210.094	3512.5
1-4	167.290	2227.0	1-8	188.888	2839.2	67.	210.487	3525.7
3-8	167.683	2237.5	1-4	189.281	2851.0	1-8	210.879	3538.8
1-2	168.075	2248.0	3-8	189.674	2862.9	1-4	211.272	3552.0
5-8	168.468	2258.5	1-2	190.066	2874.8	3-8	211.665	3565.2
3-4	168.861	2269.1	5-8	190.459	2886.6	1-2	212.058	3578.5
7-8	169.253	2279.6	3-4	190.852	2898.6	5-8	212.450	3591.7
54.	169.646	2290.2	7-8	191.214	2910.5	3-4	212.843	3605.0
1-8	170.039	2300.8	61.	191.617	2922.5	7-8	213.236	3618.3
1-4	170.431	2311.5	1-8	192.030	2934.5	68.	213.628	3631.7
3-8	170.824	2322.1	1-4	192.423	2946.5	1-8	214.021	3645.0
1-2	171.217	2332.8	3-8	192.815	2958.5	1-4	214.414	3658.4
5-8	171.609	2343.5	1-2	193.208	2970.6	3-8	214.806	3671.8
3-4	172.002	2354.3	5-8	193.601	2982.7	1-2	215.199	3685.3
7-8	172.395	2365.0	3-4	193.993	2994.8	5-8	215.592	3698.7
55.	172.788	2375.8	7-8	194.386	3006.9	3-4	215.984	3712.2
1-8	173.180	2386.6	62.	194.779	3019.1	7-8	216.377	3725.7
1-4	173.573	2397.5	1-8	195.171	3031.3	69.	216.770	3739.3
3-8	173.966	2408.3	1-4	195.564	3043.5	1-8	217.163	3752.8
1-2	174.358	2419.2	3-8	195.957	3055.7	1-4	217.555	3766.4
5-8	174.751	2430.1	1-2	196.350	3068.0	3 8	217.948	3780.0
3-4	175.144	2441.1	5-8	196.742	3080.3	1-2	218.341	3793.7
7-8	175.536	2452.0	3-4	197.135	3092.6	5-8	218.733	3807.3
56.	175.929	2463.0	7-8	197.528	3104.9	3-4	219.126	3821.0
1-8	176.322	2474.0	63.	197.920	3117.2	7-8	219.519	3834.7
1-4	176.715	2485.0	1-8	198.313	3129.6	70.	219.911	3848.5
3-8	177.107	2496.1	1-4	198.706	3142.0	1-8	220.304	3862.2
1-2	177.500	2507.2	3-8	199.098	3154.5	1-4	220.697	3876.0
5-8	177.898	2518.3	1-2	199.491	3166.9	3-8	221.090	3889.8
3-4	178.285	2529.4	5-8	199.884	3179.4	1-2	221.482	3903.6
7-8	178.678	2540.6	3-4	200.277	3191.9	5-8	221.875	3917.5
57.	179.071	2551.8	7-8	200.669	3204.4	3-4	222.268	3931.4
1-8	179.463	2563.0	64.	201.062	3217.0	7-8	222.660	3945.3
1-4	179.856	2574.2	1-8	201.455	3229.6	71.	223.053	3959.2
3-8	180.249	2585.4	1-4	201.847	3242.2	1-8	223.446	3973.1
1-2	180.642	2596.7	3-8	202.240	3254.8	1-4	223.838	3987.1
5-8	181.034	2608.0	1-2	202.633	3267.5	3-8	224.231	4001.1
3-4	181.427	2619.4	5-8	203.025	3280.1	1-2	224.624	4015.2
7-8	181.820	2630.7	3-4	203.418	3292.8	5-8	225.017	4029.2

TABLE OF CIRCLES—*Continued.*

DIAM. INS.	CIR-CUMF. INS.	AREA. SQ. INS.	DIAM. INS.	CIR-CUMF. INS.	AREA. SQ. INS.	DIAM. INS.	CIR-CUMF. INS.	AREA. SQ. INS.
71 3-4	225.409	4043.3	78 5-8	247.008	4855.2	85 1-2	268.606	5741.5
7-8	225.802	4057.4	3-4	247.400	4870.7	5-8	268.999	5758.3
72.	226.195	4071.5	7-8	247.793	4886.2	3-4	269.392	5775.1
1-8	226.587	4085.7	79.	248.186	4901.7	7-8	269.784	5791.9
1-4	226.980	4099.8	1-8	248.579	4917.2	86.	270.177	5808.8
3-8	227.373	4114.0	1-4	248.971	4932.7	1-8	270.570	5825.7
1-2	227.765	4128.2	3-8	249.364	4948.3	1-4	270.962	5842.6
5-8	228.158	4142.5	1-2	249.757	4963.9	3-8	271.355	5859.6
3-4	228.551	4156.8	5-8	250.149	4979.5	1-2	271.748	5876.5
7-8	228.944	4171.1	3-4	250.542	4995.2	5-8	272.140	5893.5
73.	229.336	4185.4	7-8	250.935	5010.9	3-4	272.533	5910.6
1-8	229.729	4199.7	80.	251.327	5026.5	7-8	272.926	5927.6
1-4	230.122	4214.1	1-8	251.720	5042.3	87.	273.319	5944.7
3-8	230.514	4228.5	1-4	252.113	5058.0	1-8	273.711	5961.8
1-2	230.907	4242.9	3-8	252.506	5073.8	1-4	274.104	5978.9
5-8	231.300	4257.4	1-2	252.898	5089.6	3-8	274.497	5996.0
3-4	231.692	4271.8	5-8	253.291	5105.4	1-2	274.889	6013.2
7-8	232.085	4286.3	3-4	253.684	5121.2	5-8	275.282	6030.4
74.	232.478	4300.8	7-8	254.076	5137.1	3-4	275.675	6047.6
1-8	232.871	4315.4	81.	254.469	5153.0	7-8	276.067	6064.9
1-4	233.263	4329.9	1-8	254.862	5168.9	88.	276.460	6082.1
3-8	233.656	4344.5	1-4	255.254	5184.9	1-8	276.853	6099.4
1-2	234.049	4359.2	3-8	255.647	5200.8	1-4	277.246	6116.7
5-8	234.441	4373.8	1-2	256.040	5216.8	3-8	277.638	6134.1
3-4	234.834	4388.5	5-8	256.433	5232.8	1-2	278.031	6151.4
7-8	235.227	4403.1	3-4	256.825	5248.8	5-8	278.424	6168.8
75.	235.619	4417.9	7-8	257.218	5264.9	3-4	278.816	6186.2
1-8	236.012	4432.6	82.	257.611	5281.0	7-8	279.209	6203.7
1-4	236.405	4447.4	1-8	258.003	5297.1	89.	279.602	6221.1
3-8	236.798	4462.2	1-4	258.396	5313.3	1-8	279.994	6238.6
1-2	237.190	4477.0	3-8	258.789	5329.4	1-4	280.387	6256.1
5-8	237.583	4491.8	1-2	259.181	5345.6	3-8	280.780	6273.7
3-4	237.976	4506.7	5-8	259.574	5361.8	1-2	281.173	6291.2
7-8	238.368	4521.5	3-4	259.967	5378.1	5-8	281.565	6308.8
76.	238.761	4536.5	7-8	260.359	5394.3	3-4	281.958	6326.4
1-8	239.154	4551.4	83.	260.752	5410.6	7-8	282.351	6344.1
1-4	239.546	4566.4	1-8	261.145	5426.9	90.	282.743	6361.7
3-8	239.939	4581.3	1-4	261.538	5443.3	1-8	283.136	6379.4
1-2	240.332	4596.3	3-8	261.930	5459.6	1-4	283.529	6397.1
5-8	240.725	4611.4	1-2	262.323	5476.0	3-8	283.921	6414.9
3-4	241.117	4626.4	5-8	262.716	5492.4	1-2	284.314	6432.6
7-8	241.510	4641.5	3-4	263.108	5508.8	5-8	284.707	6450.4
77.	241.903	4656.6	7-8	263.501	5525.3	3-4	285.100	6468.2
1-8	242.295	4671.8	84.	263.894	5541.8	7-8	285.492	6486.0
1-4	242.688	4686.9	1-8	264.286	5558.3	91.	285.885	6503.9
3-8	243.081	4702.1	1-4	264.679	5574.8	1-8	286.278	6521.8
1-2	243.473	4717.3	3-8	265.072	5591.3	1-4	286.670	6539.7
5-8	243.866	4732.5	1-2	265.465	5607.9	3-8	287.063	6557.6
3-4	244.259	4747.8	5-8	265.857	5624.5	1-2	287.456	6575.5
7-8	244.652	4763.1	3-4	266.250	5641.2	5-8	287.848	6593.5
78.	245.044	4778.4	7-8	266.643	5657.8	3-4	288.241	6611.5
1-8	245.437	4793.7	85.	267.035	5674.5	7-8	288.634	6629.6
1-4	245.830	4809.0	1-8	267.428	5691.2	92.	289.027	6647.6
3-8	246.222	4824.4	1-4	267.821	5707.9	1-8	289.419	6665.7
1-2	246.615	4839.8	3-8	268.213	5724.7	1-4	289.812	6683.8

TABLE OF CIRCLES—*Continued.*

DIAM. INS.	CIR-CUMF. INS.	AREA. SQ. INS.	DIAM. INS.	CIR-CUMF. INS.	AREA. SQ. INS.	DIAM. INS.	CIR-CUMF. INS.	AREA. SQ. INS.
92 3-8	290.205	6701.9	95.	298.451	7068.2	97 5-8	306.698	7485.3
1-2	290.597	6720.1	1-8	298.844	7106.9	3-4	307.091	7504.5
5-8	290.990	6738.2	1-4	299.237	7125.6	7-8	307.483	7523.7
3-4	291.383	6756.4	3-8	299.629	7144.3	98.	307.876	7543.0
7-8	291.775	6774.7	1-2	300.022	7163.0	1-8	308.269	7562.2
93.	292.168	6792.9	5-8	300.415	7181.8	1-4	308.661	7581.5
1-8	292.561	6811.2	3-4	300.807	7200.6	3-8	309.054	7600.8
1-4	292.954	6829.5	7-8	301.200	7219.4	1-2	309.447	7620.1
3-8	293.346	6847.8	96.	301.593	7238.2	5-8	309.840	7639.5
1-2	293.739	6866.1	1-8	301.986	7257.1	3-4	310.232	7658.9
5-8	294.132	6884.5	1-4	302.378	7276.0	7-8	310.625	7678.3
3-4	294.524	6902.9	3-8	302.771	7294.9	99.	311.018	7697.7
7-8	294.917	6921.3	1-2	303.164	7313.8	1-8	311.410	7717.1
94.	295.310	6939.8	5-8	303.556	7332.8	1-4	311.803	7736.6
1-8	295.702	6958.2	3-4	303.949	7351.8	3-8	312.196	7756.1
1-4	296.095	6976.7	7-8	304.342	7370.8	1-2	312.588	7775.6
3-8	296.488	6995.3	97.	304.734	7389.8	5-8	312.981	7795.2
1-2	296.881	7013.8	1-8	305.127	7408.9	3-4	313.374	7814.8
5-8	297.273	7032.4	1-4	305.520	7428.0	7-8	313.767	7834.4
3-4	297.666	7051.0	3-8	305.913	7447.1	100.	314.159	7854.0
7-8	298.059	7069.6	1-2	306.305	7466.2			

WEIGHT OF A LINEAL FOOT OF ROUND AND SQUARE IRON.

Size in Inches.	Rounds. Weight per foot.	Squares. Weight per foot.	Size in Inches.	Rounds. Weight per foot.	Squares. Weight per foot.
1/16	0.01	0.013	3 3/8	29.82	37.969
1/8	0.041	0.052	3 1/2	32.07	40.833
3/16	0.092	0.117	3 5/8	34.40	43.802
1/4	0.163	0.208	3 3/4	36.813	46.875
3/8	0.363	0.468	3 7/8	39.31	50.052
1/2	0.654	0.833	4	41.887	53.333
5/8	1.022	1.302	4 1/8	44.547	56 719
3/4	1.472	1.875	4 1/4	47.287	60.208
7/8	2.004	2.552	4 3/8	50.11	63.802
1	2.618	3.333	4 1/2	53.013	67.50
1 1/8	3.313	4.218	4 5/8	56.00	71.302
1 1/4	4.09	5.208	4 3/4	59.067	75.208
1 3/8	4.947	6.302	4 7/8	62.217	79.219
1 1/2	5.89	7.50	5	65.45	83.333
1 5/8	6.91	8.802	5 1/8	68.763	87.552
1 3/4	8.017	10.208	5 1/4	72.157	91.875
1 7/8	9.203	11.718	5 3/8	75.633	96.302
2	10.47	13.333	5 1/2	79.197	100.833
2 1/8	11.82	15.052	5 5/8	82.833	105.468
2 1/4	13.253	16.875	5 3/4	86.557	110.208
2 3/8	14.766	18.802	5 7/8	90.36	115.052
2 1/2	16.36	20.833	6	94.247	120.00
2 5/8	18.036	22.969	6 1/4	102.263	130.208
2 3/4	19.797	25.208	6 1/2	110.61	140.833
3	23.56	30.00	6 3/4	119.28	151.875
3 1/8	25.563	32.552	7	128.28	163.333
3 1/4	27.65	35.208			

WEIGHT OF A LINEAL FOOT OF FLAT IRON.

Width in Inches.	THICKNESS IN INCHES.											
	1/16	1/8	3/16	1/4	5/16	3/8	7/16	1/2	5/8	3/4	7/8	1
3/4	0.16	0.31	0.47	0.63	0.77	0.94	1.09	1.25	1.56	1.87	2.18	2.50
7/8	0.18	0.36	0.55	0.73	0.91	1.09	1.28	1.46	1.82	2.19	2.56	2.92
1	0.21	0.42	0.62	0.83	1.04	1.25	1.46	1.67	2.08	2.50	2.92	3.33
1 1/8	0.23	0.47	0.70	0.94	1.17	1.41	1.64	1.88	2.34	2.81	3.28	3.75
1 1/4	0.26	0.52	0.78	1.04	1.30	1.56	1.82	2.08	2.60	3.12	3.64	4.17
1 3/8	0.29	0.57	0.86	1.15	1.43	1.72	2.00	2.29	2.86	3.44	4.01	4.58
1 1/2	0.31	0.63	0.94	1.25	1.56	1.88	2.19	2.50	3.18	3.75	4.38	5.00
1 5/8	0.34	0.68	1.02	1.35	1.69	2.03	2.37	2.71	3.38	4.06	4.74	5.42
1 3/4	0.36	0.73	1.09	1.46	1.82	2.19	2.55	2.92	3.65	4.37	5.10	5.83
1 7/8	0.39	0.78	1.17	1.56	1.95	2.34	2.73	3.12	3.91	4.68	5.46	6.25
2	0.42	0.83	1.24	1.67	2.08	2.50	2.92	3.33	4.17	5.00	5.83	6.67
2 1/8	0.44	0.89	1.33	1.77	2.21	2.66	3.10	3.54	4.43	5.31	6.20	7.08
2 1/4	0.47	0.94	1.41	1.88	2.34	2.81	3.28	3.75	4.69	5.63	6.56	7.50
2 3/8	0.49	0.99	1.48	1.98	2.47	2.96	3.46	3.96	4.95	5.94	6.93	7.92
2 1/2	0.52	1.04	1.56	2.08	2.60	3.12	3.64	4.17	5.21	6.25	7.29	8.33
2 5/8	0.55	1.09	1.64	2.19	2.73	3.28	3.83	4.38	5.47	6.56	7.66	8.75
2 3/4	0.57	1.14	1.72	2.29	2.86	3.44	4.01	4.59	5.73	6.87	8.02	9.17
2 7/8	0.60	1.20	1.80	2.40	2.99	3.59	4.19	4.79	5.99	7.19	8.38	9.58
3	0.62	1.25	1.87	2.50	3.12	3.75	4.37	5.00	6.25	7.50	8.75	10.00
3 1/4	0.68	1.35	2.03	2.71	3.38	4.07	4.74	5.42	6.77	8.12	9.48	10.83
3 1/2	0.73	1.46	2.19	2.92	3.65	4.38	5.11	5.83	7.29	8.75	10.21	11.67
3 3/4	0.78	1.56	2.34	3.12	3.90	4.69	5.47	6.25	7.81	9.37	10.94	12.50
4	0.83	1.67	2.50	3.33	4.17	5.00	5.83	6.67	8.33	10.00	11.67	13.33
4 1/2	0.94	1.87	2.81	3.75	4.69	5.63	6.56	7.50	9.38	11.25	13.13	15.00
5	1.04	2.08	3.13	4.17	5.21	6.25	7.30	8.34	10.42	12.50	14.59	16.67
6	1.25	2.50	3.75	5.00	6.25	7.50	8.75	10.00	12.50	15.00	17.50	20.00
7	1.46	2.92	4.37	5.83	7.29	8.75	10.20	11.67	14.58	17.50	20.42	23.33
8	1.67	3.33	5.00	6.67	8.34	10.00	11.67	13.33	16.67	20.00	23.33	26.67
9	1.87	3.75	5.62	7.50	9.37	11.25	13.12	15.00	18.75	22.50	26.25	30.00
10	2.08	4.17	6.25	8.33	10.42	12.50	14.58	16.67	20.83	25.00	29.17	33.33
11	2.29	4.58	6.87	9.17	11.46	13.75	16.04	18.33	22.92	27.50	32.08	36.67
12	2.50	5.00	7.50	10.00	12.50	15.00	17.50	20.00	25.00	30.00	35.00	40.00

DECIMAL EQUIVALENTS FOR FRACTIONS OF AN INCH.

FRACTION.	DECIMAL.	FRACTION.	DECIMAL.
$\frac{1}{64}$	.015625	$\frac{33}{64}$	.515625
$\frac{1}{32}$	.03125	$\frac{17}{32}$	.53125
$\frac{3}{64}$	.046875	$\frac{35}{64}$	.546875
$\frac{1}{16}$	.0625	$\frac{9}{16}$	.5625
$\frac{5}{64}$	.078125	$\frac{37}{64}$	.578125
$\frac{3}{32}$	.09375	$\frac{19}{32}$	.59375
$\frac{7}{64}$	.109375	$\frac{39}{64}$	.609375
$\frac{1}{8}$	.125	$\frac{5}{8}$	.625
$\frac{9}{64}$	.140625	$\frac{41}{64}$	.640625
$\frac{5}{32}$	.15625	$\frac{21}{32}$	.65625
$\frac{11}{64}$	.171875	$\frac{43}{64}$	.671875
$\frac{3}{16}$	.1875	$\frac{11}{16}$	.6875
$\frac{13}{64}$	.203125	$\frac{45}{64}$	.703125
$\frac{7}{32}$	.21875	$\frac{23}{32}$	.71875
$\frac{15}{64}$	.234375	$\frac{47}{64}$	.734375
$\frac{1}{4}$	.25	$\frac{3}{4}$	.75
$\frac{17}{64}$	.265625	$\frac{49}{64}$	.765625
$\frac{9}{32}$	.28125	$\frac{25}{32}$	.78125
$\frac{19}{64}$	.296875	$\frac{51}{64}$	.796875
$\frac{5}{16}$	.3125	$\frac{13}{16}$	.8125
$\frac{21}{64}$	.328125	$\frac{53}{64}$	.828125
$\frac{11}{32}$	.34375	$\frac{27}{32}$	.84375
$\frac{23}{64}$	.359375	$\frac{55}{64}$	.859375
$\frac{3}{8}$	.375	$\frac{7}{8}$	.875
$\frac{25}{64}$	.390625	$\frac{57}{64}$	.890625
$\frac{13}{32}$	.40625	$\frac{29}{32}$	.90625
$\frac{27}{64}$	.421875	$\frac{59}{64}$	.921875
$\frac{7}{16}$	.4375	$\frac{15}{16}$	.9375
$\frac{29}{64}$	.453125	$\frac{61}{64}$	.953125
$\frac{15}{32}$	.46875	$\frac{31}{32}$	.96875
$\frac{31}{64}$	.484375	$\frac{63}{64}$	.984375
$\frac{1}{2}$	.5		

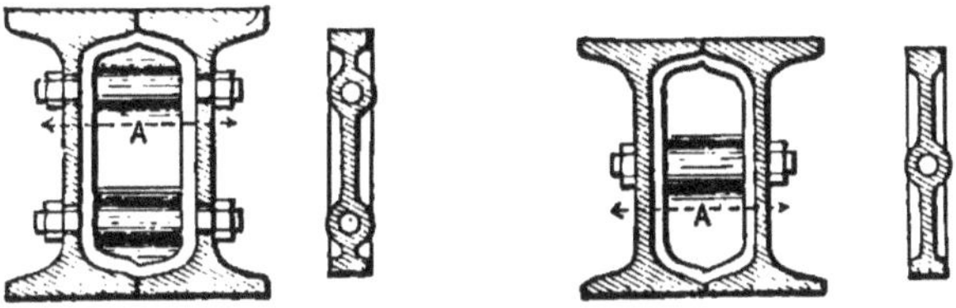

STANDARD SEPARATORS FOR PENCOYD I BEAMS.

Chart No.	Size of Beam.	Weight of separator.	Weight of each additional inch of width.	Bolts, A. No.	Bolts, A. Size.	Weight of each complete bolt.	Weight per additional inch of length.
1	15 " Heavy	22	3.84	2	¾"	1.75	.123
2	15 " Light	21	3.13	2	¾"	1.62	.123
3	12 " Heavy	16	2.76	2	¾"	1.69	.123
4	12 " Light	14½	2.95	2	¾"	1.58	.123
5	10½" Heavy	11¼	2.10	1	¾"	1.64	.123
5½	10½" Medium	11	2.06	1	¾"	1.28	.123
6	10½" Light	11	2.03	1	¾"	1.53	.123
7	10 " Heavy	10	1.93	1	¾"	1.56	.123
8	10 " Light	10	1.93	1	¾"	1.52	.123
9	9 " Heavy	9¼	1.63	1	¾"	1.52	.123
10	9 " Light	9	1.63	1	¾"	1.48	.123
11	8 " Heavy	6¾	1.36	1	¾"	1.50	.123
12	8 " Light	6¼	1.49	1	¾"	1.46	.123
13	7 " Heavy	4	1.26	1	⅝"	0.96	.085
14	7 " Light	4	1.26	1	⅝"	0.91	.085
15	6 " Heavy	3	1.24	1	⅝"	0.90	.085
16	6 " Light	3	1.24	1	⅝"	0.87	.085
17	5 " Heavy	2¾	1.10	1	½"	0.43	.055
18	5 " Light	2¾	1.10	1	½"	0.42	.055
19	4 " Heavy	2	0.85	1	½"	0.42	.055
20	4 " Light	2	0.85	1	½"	0.39	.055
21	3 " Heavy	1½	0.69	1	½"	0.38	.055
22	3 " Light	1½	0.69	1	½"	0.31	.055

The figures in the third column are the weights in lbs. for cast iron separators suitable for beams, placed with flanges in contact. When the flanges are separated, add the amount corresponding to the distance of separation, given in the fourth column. In the same way the weight of bolts may be obtained in the final columns.

Example.—A pair of 12" heavy beams have the flanges separated 1½ inches, the weight of one separator will be 2.76 × 1½ + 16 = 20.14 lbs. One ¾ bolt complete and suitable for close flanges, will weigh 1.69 lbs. Add to this .123 × 1½ = 1.88, which is the weight of bolt required.

BOLTS AND NUTS.

MANUFACTURER'S STANDARD.

Size of Nut.			Diameter of Bolt.	Weight of Head and Nut.		Weight of Bolt Bodies per Inch of Length.
Width.	Thick-ness.	Hole.		Square.	Hexagon.	
1/2	1/4	7/32	1/4	.034	.031	.014
5/8	5/16	9/32	5/16	.067	.055	.021
3/4	3/8	11/32	3/8	.110	.105	.031
7/8	7/16	13/32	7/16	.181	.171	.042
7/8	1/2	7/16	1/2	.210	.192	.055
1	1/2	7/16	1/2	.280	.233	.055
1 1/8	9/16	1/2	9/16	.369	.335	.069
1 1/8	5/8	9/16	5/8	.431	.403	.085
1 1/4	5/8	9/16	5/8	.545	.475	.085
1 1/4	3/4	9/16	5/8	.....	.563	.085
1 3/8	3/4	21/32	3/4	.776	.673	.123
1 3/8	7/8	21/32	3/4	.....	.770	.123
1 1/2	7/8	21/32	3/4	.....	.964	.123
1 5/8	7/8	25/32	7/8	1.34	1.14	.167
1 1/2	1	25/32	7/8	.....	1.19	.167
1 5/8	1	25/32	7/8	.....	1.28	.167
1 3/4	7/8	25/32	7/8	1.46	.....	.167
1 3/4	1	7/8	1	1.75	1.48	.218
1 3/4	1 1/8	7/8	1	.....	1.65	.218
2	1	7/8	1	2.24	.....	.218
2	1 1/8	15/16	1 1/8	2.47	.....	.276
2	1 1/4	15/16	1 1/8	.....	2.48	.276
2 1/8	1 1/8	15/16	1 1/8	3.14	.....	.276
2 1/4	1 1/4	1 1/16	1 1/4	3.74	.....	.341
2 1/4	1 3/8	1 1/16	1 1/4	.....	3.46	.341
2 1/2	1 1/4	1 1/16	1 1/4	4.47	.....	.341
2 1/2	1 1/2	1 3/16	1 3/8	.....	4.63	.412
2 3/4	1 3/8	1 3/16	1 3/8	5.85	.....	.412
2 3/4	1 5/8	1 5/16	1 1/2	.....	6.11	.491
3	1 1/2	1 5/16	1 1/2	7.59	.....	.491

BOLTS AND NUTS.

MANUFACTURER'S STANDARD.

Size of Nut.			Diameter of Bolt.	Weight of Head and Nut.		Weight of Bolt Bodies per inch of Length.
Width.	Thickness.	Hole.		Square.	Hexagon.	
3	$1\frac{3}{4}$	$1\frac{7}{16}$	$1\frac{5}{8}$	....	7.65	.576
$3\frac{1}{4}$	$1\frac{5}{8}$	$1\frac{7}{16}$	$1\frac{5}{8}$	9.48	....	.576
$3\frac{1}{4}$	$1\frac{7}{8}$	$1\frac{9}{16}$	$1\frac{3}{4}$	....	9.42	.668
$3\frac{1}{2}$	$1\frac{3}{4}$	$1\frac{9}{16}$	$1\frac{3}{4}$	11.9	....	.668
$3\frac{1}{2}$	2	$1\frac{11}{16}$	$1\frac{7}{8}$	....	11.6	.767
$3\frac{3}{4}$	$1\frac{7}{8}$	$1\frac{11}{16}$	$1\frac{7}{8}$	14.1	....	.767
$3\frac{1}{2}$	2	$1\frac{13}{16}$	2	....	12.0	.872
$3\frac{1}{2}$	$2\frac{1}{8}$	$1\frac{13}{16}$	2	....	12.6	.872
4	2	$1\frac{13}{16}$	2	18.6	....	.872
4	$2\frac{1}{8}$	$1\frac{7}{8}$	$2\frac{1}{8}$	18.9	....	.985
4	$2\frac{1}{4}$	2	$2\frac{1}{4}$	19.3	....	1.104

The preceding tables for bolts and nuts include the sizes of nuts usually applied to structural work.

The sizes known as "U. S." or "Franklin Inst." standard, used on finished machines, are lighter than the foregoing.

The weights given in the fifth and sixth columns are for a head and nut, or for two nuts, including the portion of the bolt body contained in the nuts.

The final column is the weight of bolt bodies of round iron per inch of length. To obtain the weight of any bolt: multiply the amount in final column by the clear length between nuts in inches and add in the weight of nuts as given in the fifth or sixth column.

Example.—What is the weight of a bolt $\frac{3}{4}''$ diam. and 20″ long between nuts, the nuts being $1\frac{3}{8}''$ sq. × $\frac{3}{4}''$? .123 × 20 = 2.46 + .77 = 3.23 lbs.

WEIGHT OF BRIDGE RIVETS IN POUNDS.

DIAMETER OF RIVET.	WEIGHT OF TWO HEADS.	WEIGHT OF BODY PER INCH OF LENGTH.
$\frac{3}{8}$	.036	.031
$\frac{7}{16}$	.058	.0J2
$\frac{1}{2}$	.080	.054
$\frac{9}{16}$	.120	.069
$\frac{5}{8}$	.160	.085
$1\frac{1}{16}$	.210	.103
$\frac{3}{4}$	.260	.123
$1\frac{3}{16}$	.350	.144
$\frac{7}{8}$	.440	.167
$1\frac{8}{16}$	.540	.192
1	.640	.218
$1\frac{1}{16}$	.714	.246
$1\frac{1}{8}$	.788	276
$1\frac{1}{4}$	1.07	.341

This table applies to rivets whose heads are a spherical segment, the contents being equal to a hemisphere whose diameter equals $1\frac{1}{2}$ diameters of the rivet shank plus $\frac{1}{16}$th of an inch. To find the weight of rivets, take the total thickness of material to be riveted, which will be the length of the rivet between heads, and multiply this by the weight per inch of length of rivet shank, and add in the weight of the two heads.

Example.—Three $\frac{3}{4}$ inch plates are to be riveted together with $\frac{7}{8}$ inch rivets, required the weight of each rivet. The length of rivet shank equals $3 \times \frac{3}{4} = 2\frac{1}{4}$ inches. Then $2\frac{1}{4} \times .167 = .376$, to which add the weight of the heads .44, making .816 lb. for each rivet.

INDEX.

SECTIONS

OF

PENCOYD SHAPES

Plate 1 Scale ¼ Size

Plates 2 to 28 Scale ⅓ Size

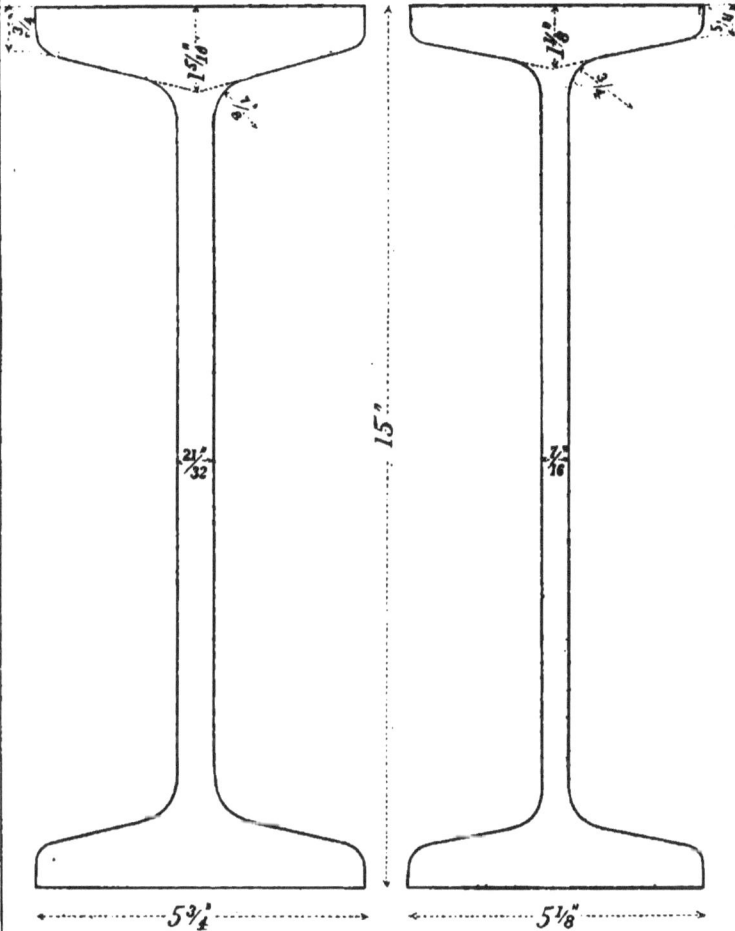

No. 1
Wt. 200 to 233 Lbs.

No. 2
Wt. 145 to 201 Lbs.

All weights given in pounds per Yard.

Wt. 168 to 194 Lbs.

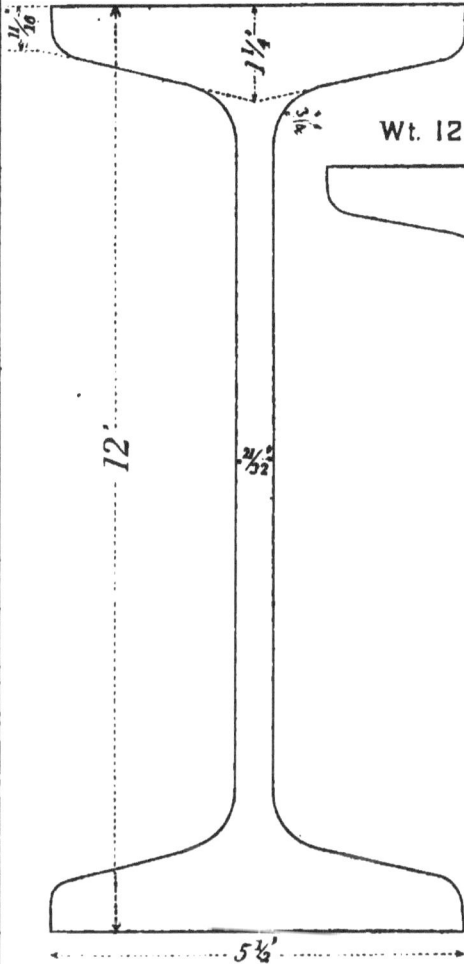

$\frac{11}{16}$"

$1\frac{1}{4}$"

$\frac{3}{4}$"

Wt. 120 to 163 Lbs.

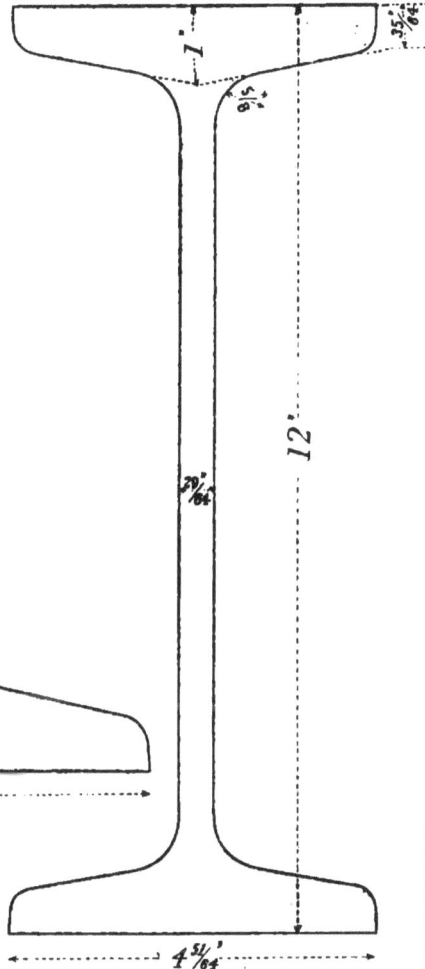

$\frac{35}{64}$"

1"

$\frac{6\frac{1}{2}}{8}$"

12'

$\frac{21}{32}$"

$\frac{29}{64}$"

12'

$5\frac{1}{2}$'

No. 3

$4\frac{51}{64}$'

No. 4

All weights given in pounds per Yard.

All weights given in pounds per Yard.

Wt. 134. to 161. Lbs.

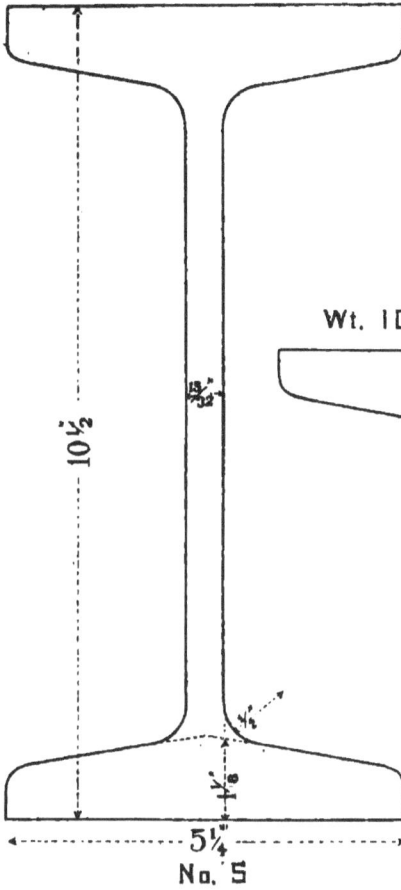

Wt. 108. to 135. Lbs.

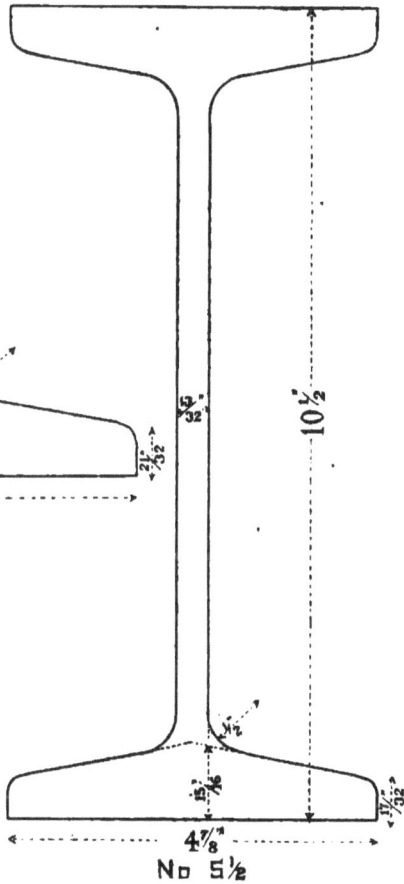

$10\frac{1}{2}$

$\frac{15}{32}$"

$10\frac{1}{2}$

$\frac{5}{32}$"

$1\frac{1}{8}$

$\frac{21}{32}$

$5\frac{1}{4}$"

No. 5

$\frac{15}{16}$

$\frac{11}{32}$"

$4\frac{7}{8}$"

No 5½

Wt. 89. to 109. Lbs.

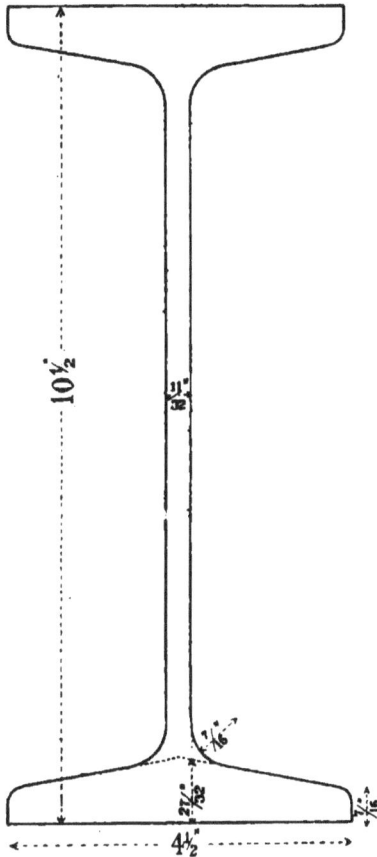

No. 6

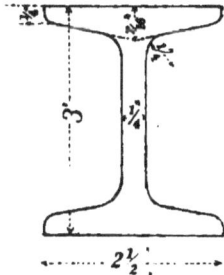

No. 21
Wt. 23 to 29 Lbs.

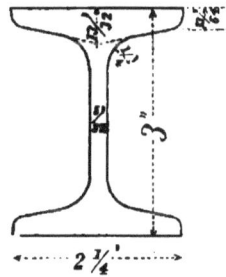

No. 22
Wt. 17 to 22 Lbs.

All weights given in pounds per Yard.

Wt. 112 to 137. Lbs.

10'

1¼'

⅜'

½'

½'

4⅝'

No. 7

Wt. 90 to 106 Lbs.

10'

³¹/₃₂'

³/₃₂'

⁵/₃₂'

³²/₃₂'

4⅜'

No. 8

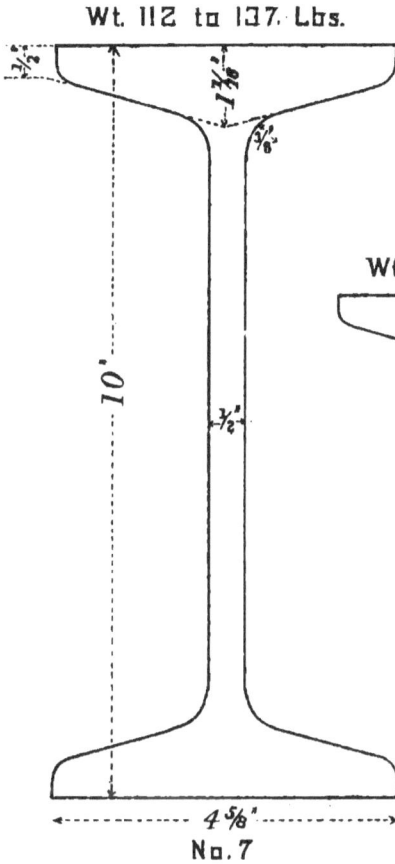

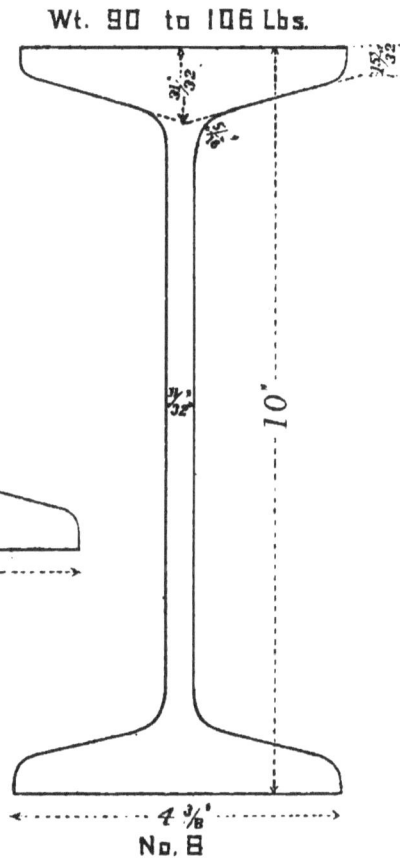

All weights given in pounds per Yard.

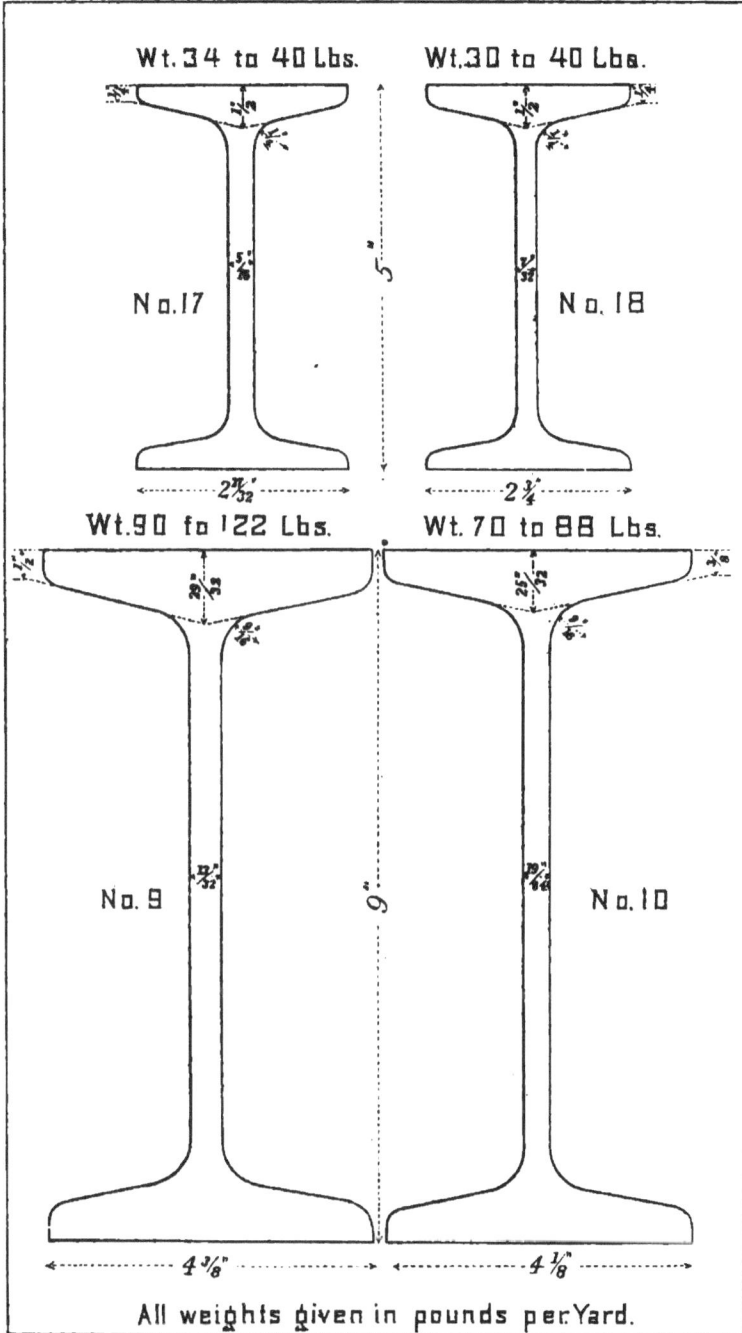

Wt. 34 to 40 Lbs.

Wt. 30 to 40 Lbs.

No. 17

No. 18

5"

2 7/32"

2 3/4"

Wt. 90 to 122 Lbs.

Wt. 70 to 88 Lbs.

No. 9

No. 10

9"

4 3/8"

4 1/8"

All weights given in pounds per Yard.

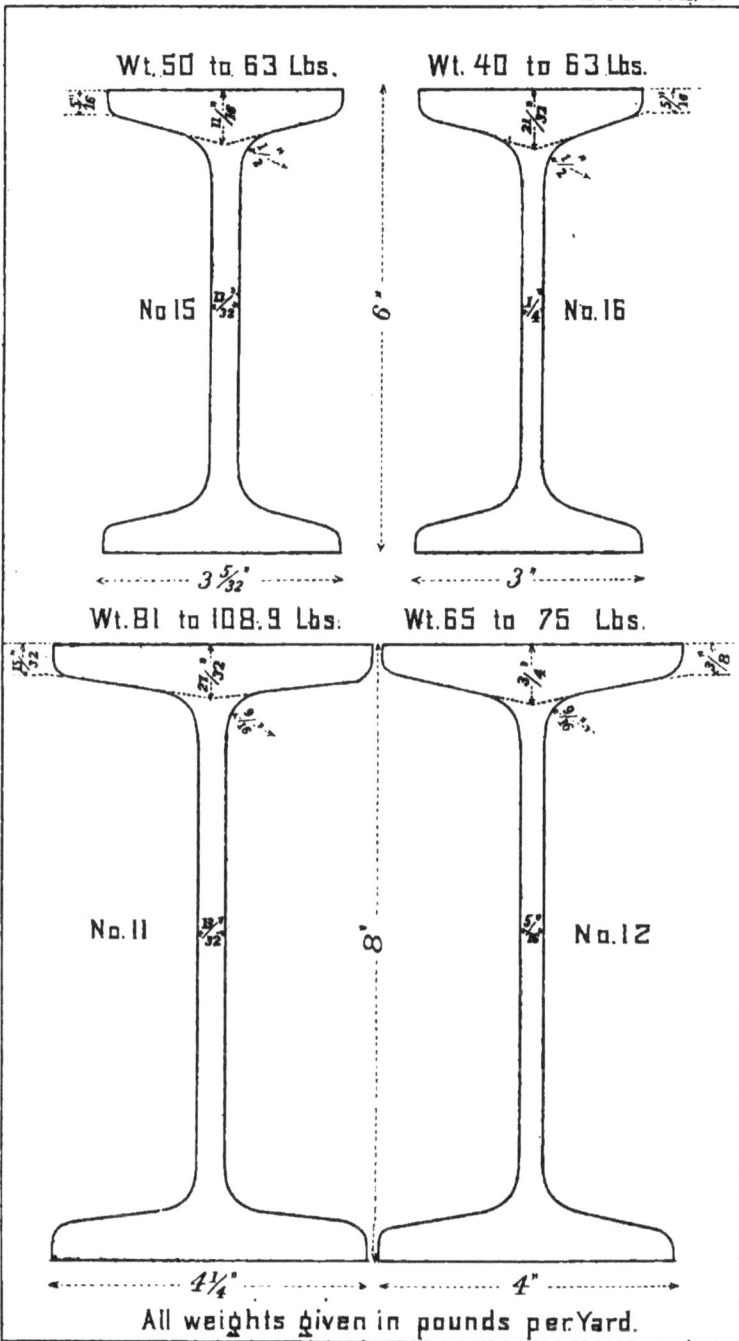

Wt. 50 to 63 Lbs.

Wt. 40 to 63 Lbs.

No. 15

No. 16

6"

3 5/32"

3"

Wt. 81 to 108.9 Lbs.

Wt. 65 to 75 Lbs.

No. 11

No. 12

8'

4 1/4"

4"

All weights given in pounds per Yard.

No.13
Wt. 65 to 88 Lbs.

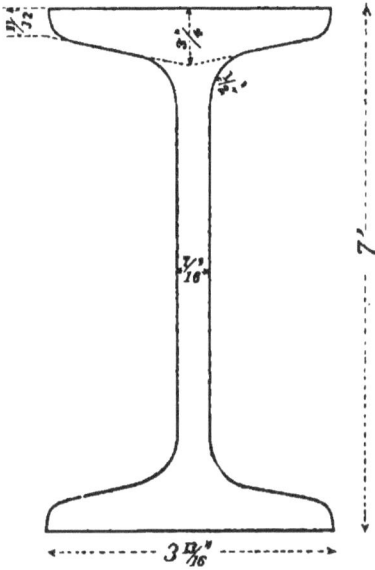

No.14
Wt. 51 to 88 Lbs.

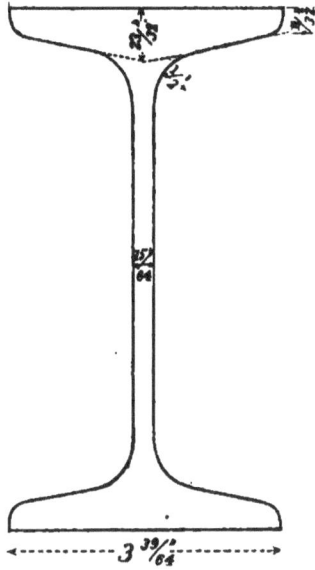

No.19
Wt. 28 to 38 Lbs.

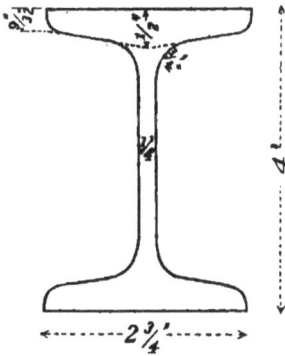

No.20
Wt. 18.5 to 21.5 Lbs.

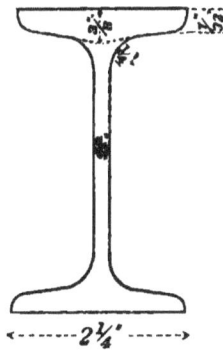

All weights given in pounds per Yard.

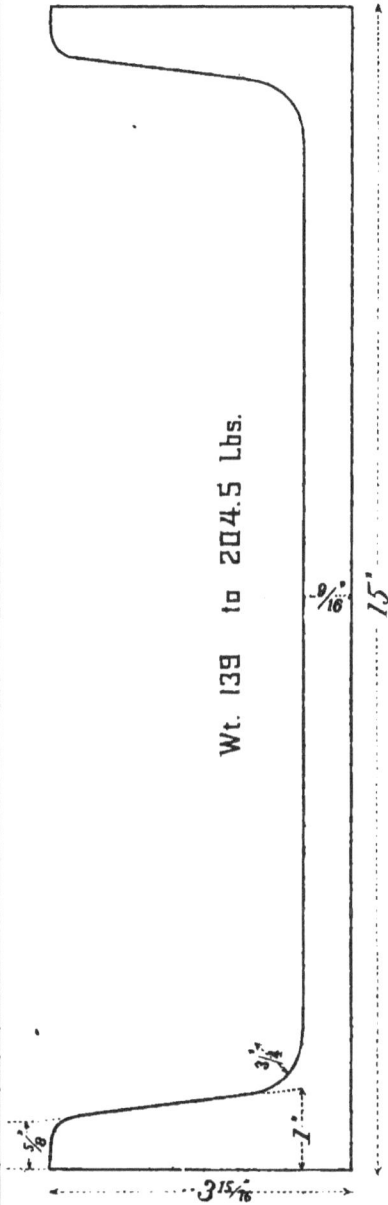

No. 30

Wt. 139 to 204.5 Lbs.

$\frac{9}{16}$"

15'

$\frac{3}{4}$"

$\frac{5}{8}$"

1'

$3\frac{15}{16}$"

No. 33

$2\frac{1}{2}$'

Car builders Channel Iron Wt. 50 to 55.

$\frac{9}{32}$"

10½'

$3\frac{1}{8}$"

$3\frac{2}{8}$"

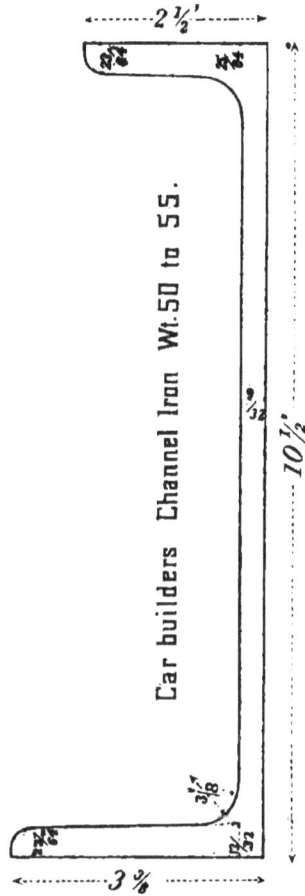

All weights given in pounds per Yard.

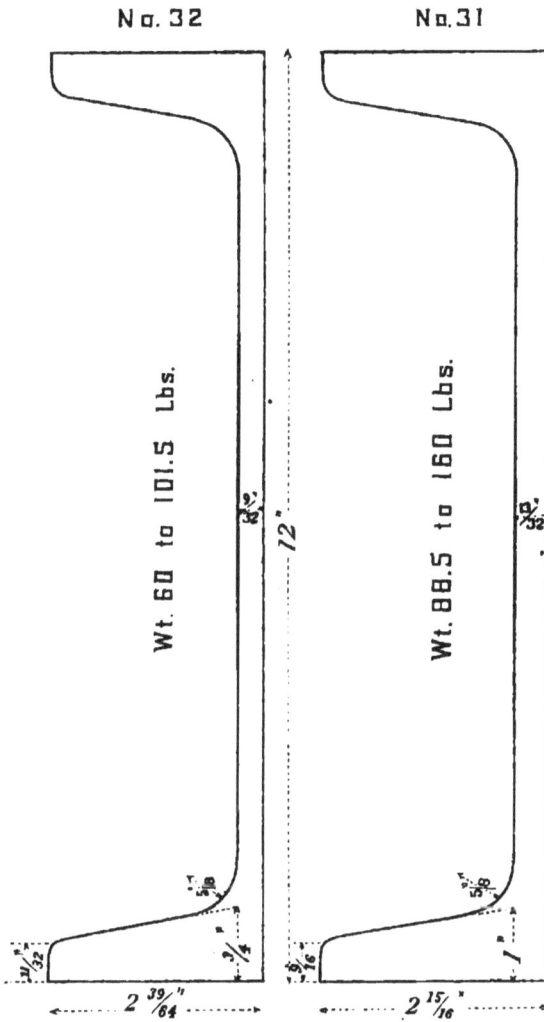

No. 32

No. 31

Wt. 60 to 101.5 Lbs.

Wt. 88.5 to 160 Lbs.

12"

All weights given in pounds per Yard.

All weights given in pounds per Yard.

No. 35　　　　No. 34

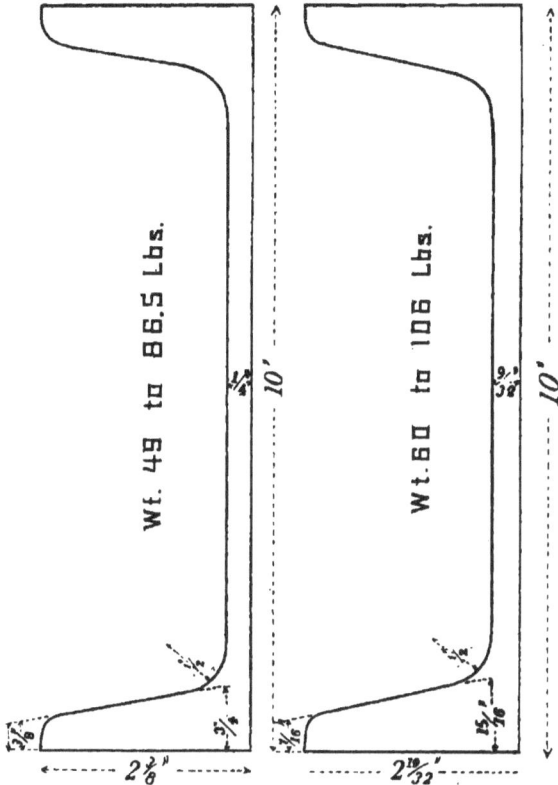

Wt. 49 to 86.5 Lbs.

Wt. 60 to 106 Lbs.

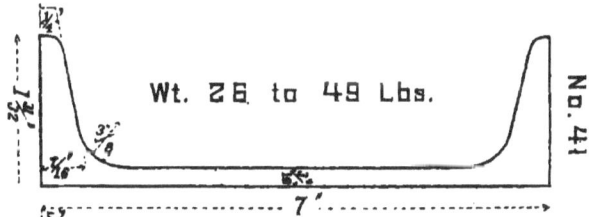

Wt. 26 to 49 Lbs.

No. 41

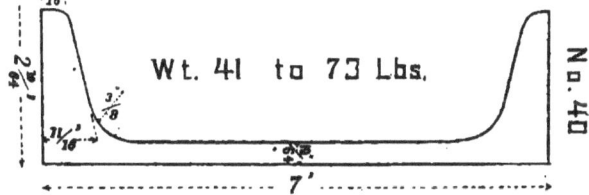

Wt. 41 to 73 Lbs.

No. 40

No. 36

No. 37

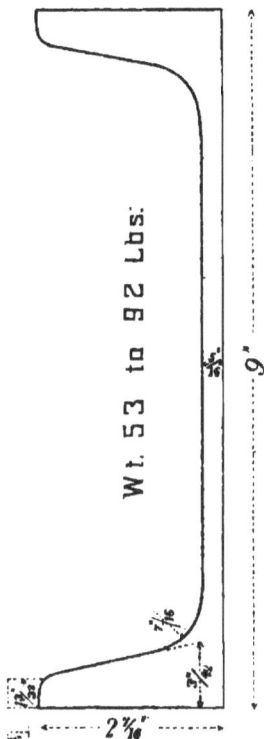

Wt. 53 to 26 Lbs.

9"

7/16

3/32

2 3/16"

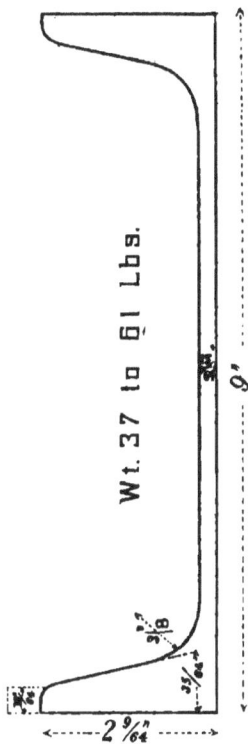

Wt. 37 to 61 Lbs.

9"

3/8

3/64

2 9/64"

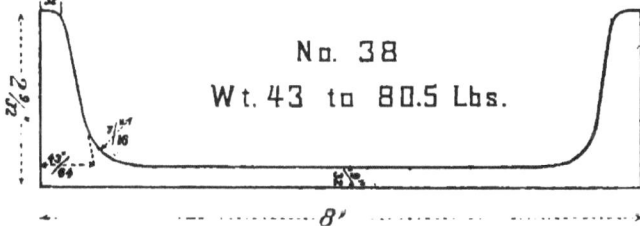

No. 38
Wt. 43 to 80.5 Lbs.

2 9/32"

7/16

13/64

8"

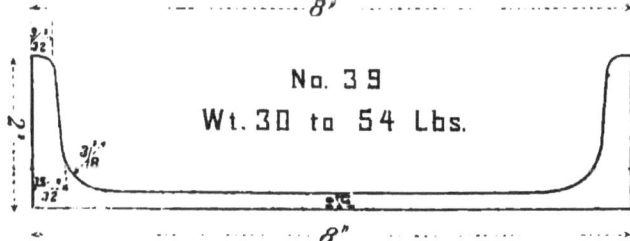

No. 39
Wt. 30 to 54 Lbs.

2"

3/16

15/32

8"

All weights given in pounds per Yard.

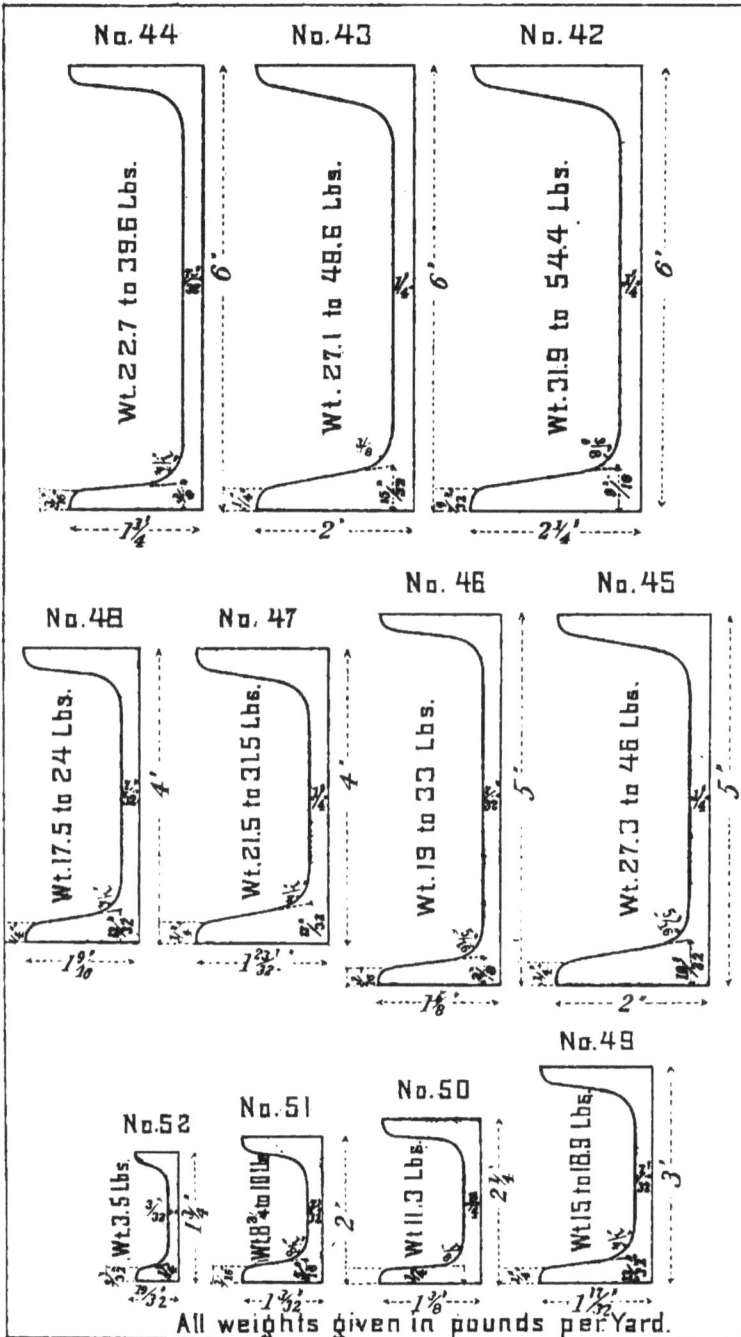

No.44   No.43   No.42

Wt. 22.7 to 39.6 Lbs.   Wt. 27.1 to 48.6 Lbs.   Wt. 31.9 to 54.4 Lbs.

6'   6'   6'

1¾'   2'   2¼'

No.48   No.47   No.46   No.45

Wt.17.5 to 24 Lbs.   Wt.21.5 to 31.5 Lbs.   Wt.19 to 33 Lbs.   Wt.27.3 to 46 Lbs.

4'   4'   5'   5'

1⁹⁄₁₀'   1²¹⁄₃₂'   1⅞'   2'

No.49

No.52   No.51   No.50

Wt.3.5 Lbs.   Wt.8¾ to 10 Lbs.   Wt.11.3 Lbs.   Wt.15 to 18.9 Lbs.

1½'   2'   2¼'   3'

¹⁹⁄₃₂'   1³⁄₃₂'   1⅛'   1¹¹⁄₁₂'

All weights given in pounds per Yard.

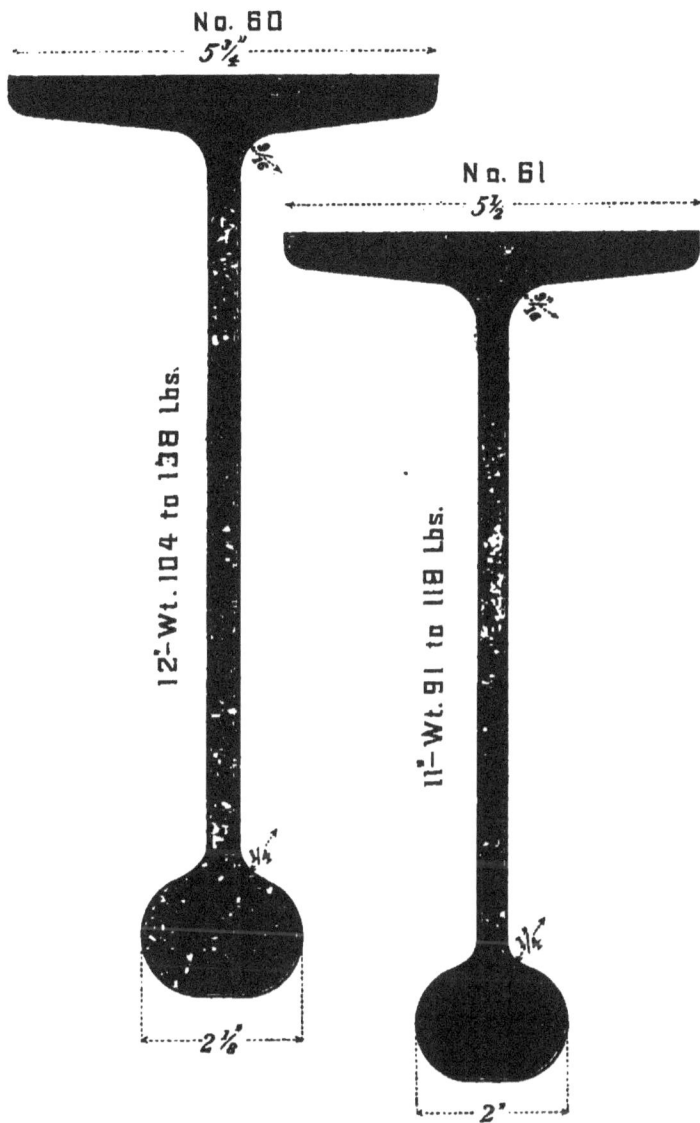

No. 60
5¾"

No. 61
5½

12'- Wt. 104 to 138 Lbs.

11"- Wt. 91 to 118 Lbs.

4½

2⅛"

3½

2"

All weights given in pounds per Yard.

No. 62

$5\frac{1}{4}'$

No. 63

5

No. 68

10"—Wt. 80 to 105 Lbs.

9"—Wt. 72 to 94 Lbs.

Wt. 62 Lbs. per Yard.

$\frac{5}{16}'$

$\frac{5}{16}'$

$1\frac{1}{32}''$

$1\frac{7}{8}'$

$1\frac{1}{32}''$

$1\frac{7}{32}'$

10'

$1\frac{7}{32}'$

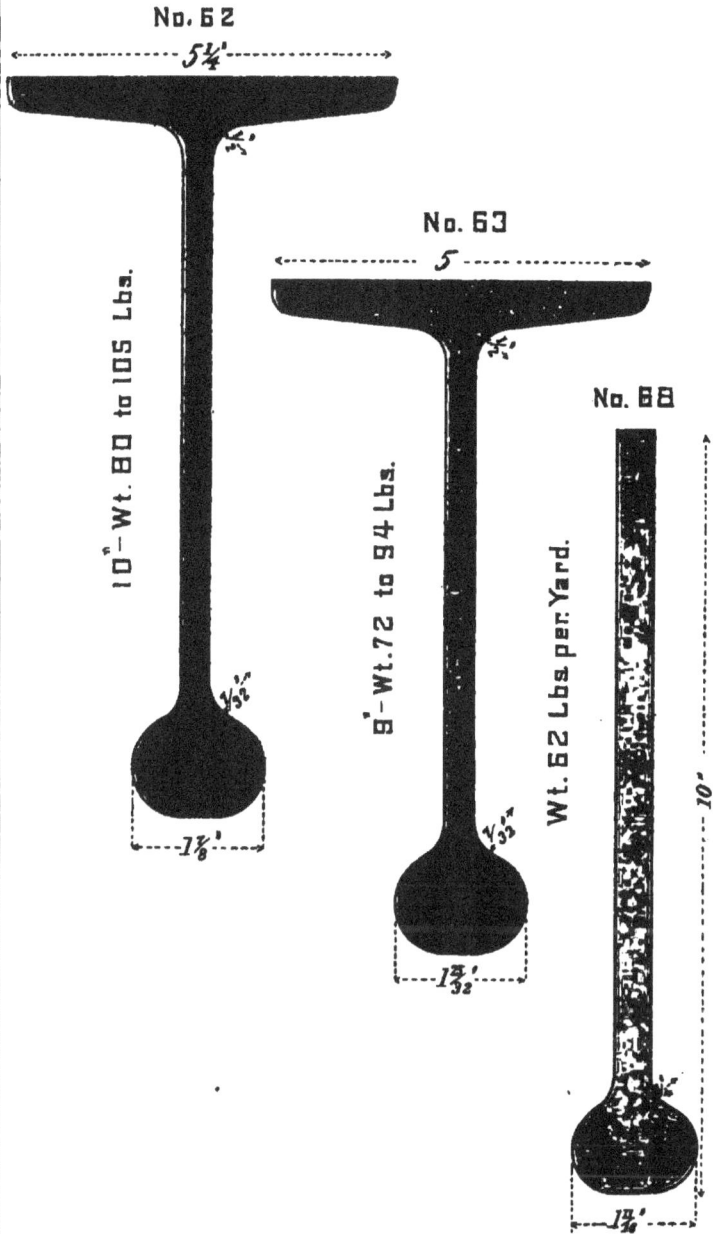

All weights given in pounds per Yard.

No. 64
4⅝"
8"—Wt. 61 to 84 Lbs.

No. 65
4¼"
7"—Wt. 52 to 72 Lbs.

No. 66
3¾"
6"—Wt. 42 to 57 Lbs.

No. 67
3¼"
5"—Wt. 34 to 46 Lbs.

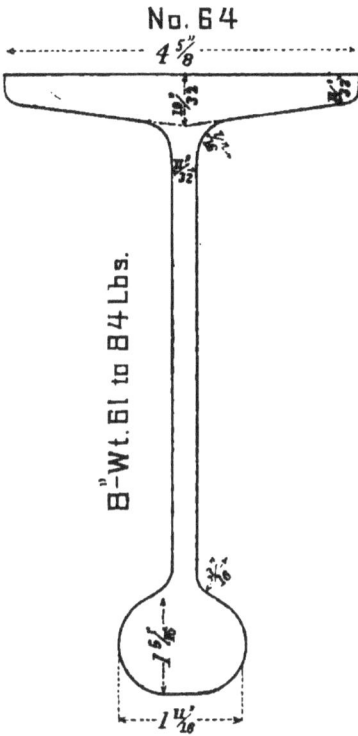

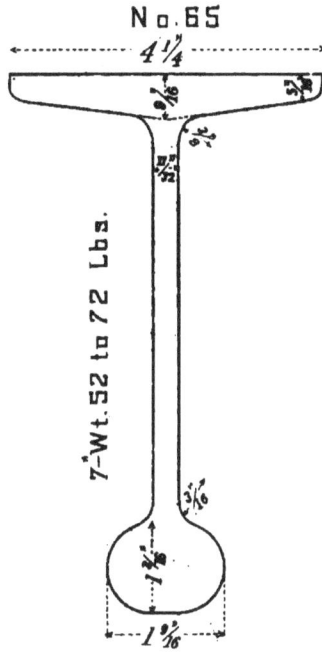

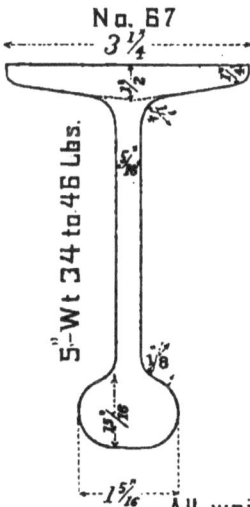

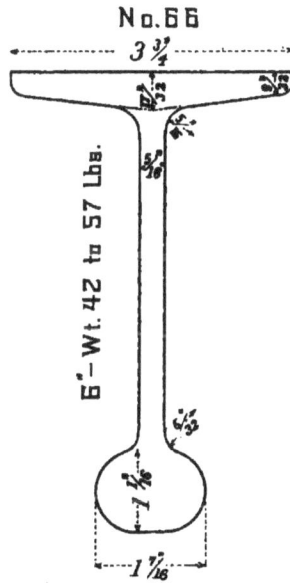

All weights given in pounds per Yard.

Wt. 36.5 Lbs.
4"
No.70

Wt. 31 Lbs.
3½"
3½"
No.71

Wt. 3 Lbs.
1"
1"
No.81

Wt. 26. Lbs.
3"
3"
No.72

Wt.4.5 Lbs.
1¼"
1¼"
No.80

Wt. 19.5 Lbs.
2½"
2½"
No.73

Wt. 17. 52 Lbs.
2½"
2½"
No.74

Wt. 6 Lbs.
1½"
1½"
No.79

Wt. 11.75 Lbs.
2¼"
2¼"
No.75

Wt. 12 Lbs.
2¼"
2¼"
No.76

Wt. 7.1 Lbs.
1¾"
1¾"
No.78

Wt. 10.5 Lbs.
2"
2"
No.77

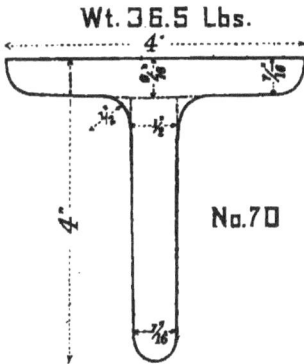

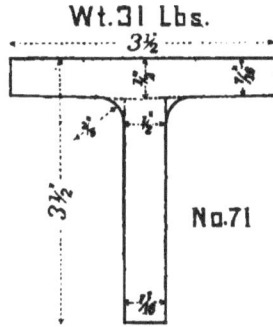

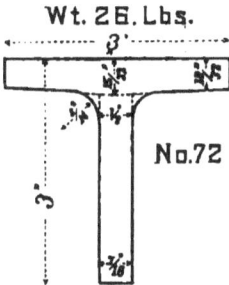

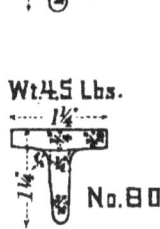

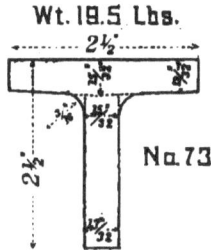

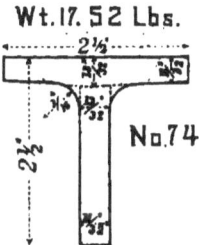

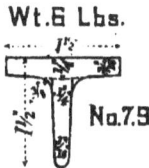

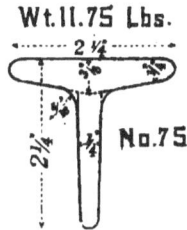

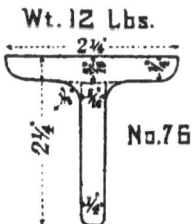

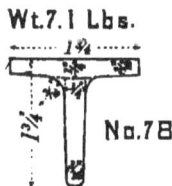

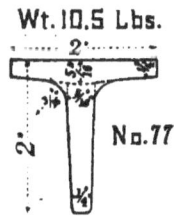

All weights given in pounds per Yard.

Wt. 22.6 Lbs.

No. 83

Wt. 19.3 Lbs.

No. 82

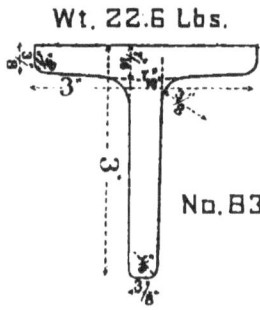

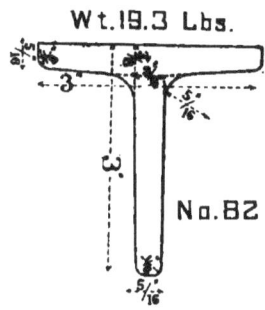

All weights given in pounds per Yard.

Wt. 44.1 Lbs.

5'

19/32"

3/8"

No. 107

4"

1/2"

Wt. 20.4 Lbs.

4'

5/16"

3/16"

2"

No. 96

3/8"

Wt. 48, 44 Lbs.

5'

1/2"

7/8"

3½"

No. 106

11/16"

Wt. 28.25 Lbs.

3'

3/8"

1/2"

1/2"

3½"

No. 97

3/16"

Wt. 11.2 Lbs.

3'

11/32"

1/4"

1½"

No. 99

3/8"

Wt. 23.75 Lbs.

3'

3/8"

3/8"

1/8"

2½"

No. 98

1/2"

Wt. 18, 75 Lbs.

2¾'

3/8"

3/4"

1¾"

No. 104

1/4"

Wt. 21. Lbs.

2¾'

3/8"

3/4"

2"

No. 105

1/4"

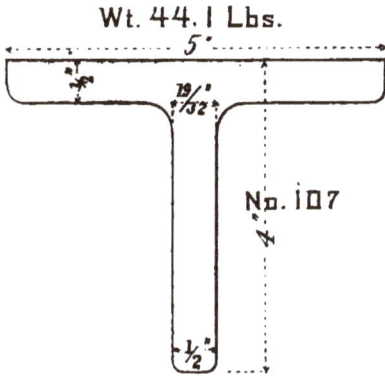

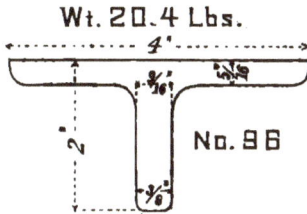

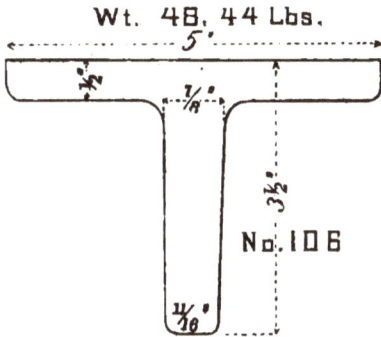

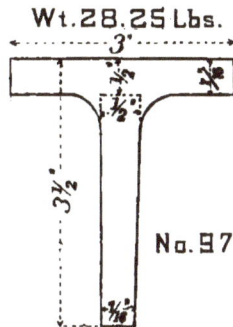

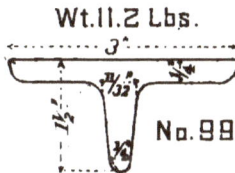

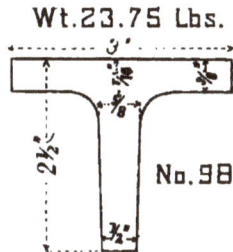

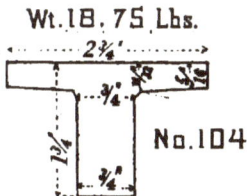

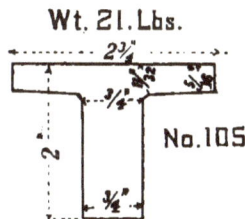

All weights given in pounds per Yard.

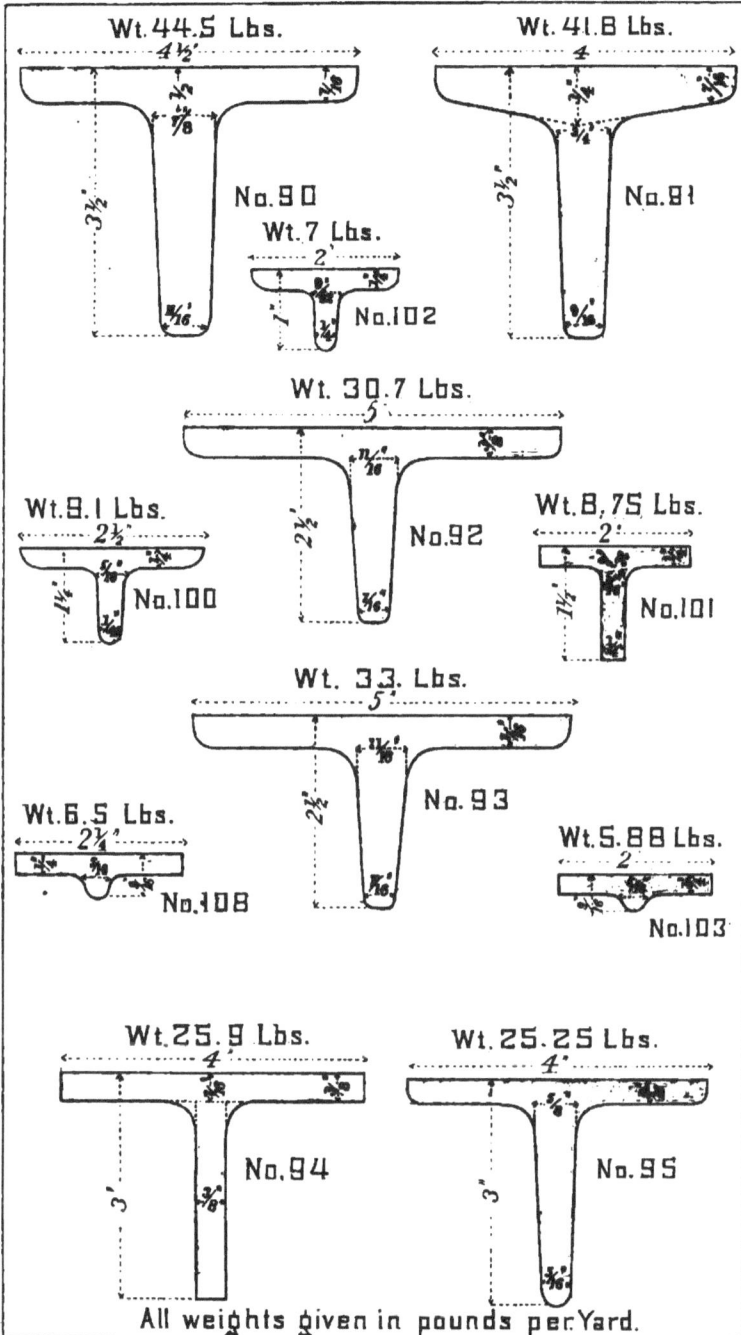

All weights given in pounds per Yard.

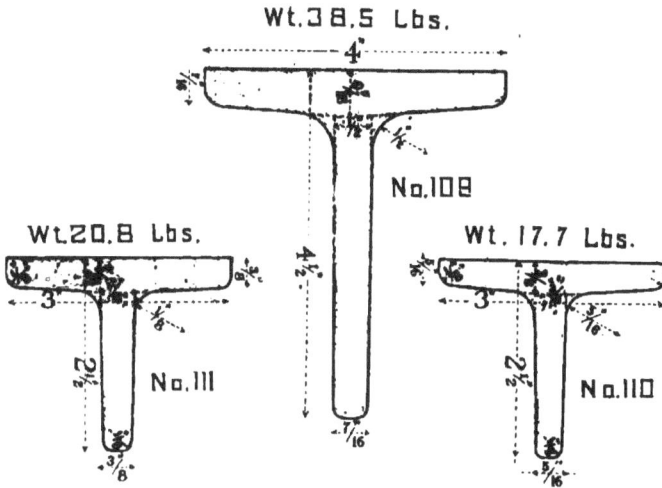

Wt. 38.5 Lbs.

No. 108

Wt. 20.8 Lbs.

No. 111

Wt. 17.7 Lbs.

No. 110

All weights given in pounds per Yard.

6" × 6" 50.6 to 110 Lbs. $\frac{5}{16}$

No.120

5" × 5" 41.8 Lbs to 90 Lbs. $\frac{7}{16}$

No.121

4" × 4" 28.8 to 54.4 Lbs. $\frac{3}{16}$

No.122

$3\frac{1}{2}$ × $3\frac{1}{2}$ 21 to 39.8 Lbs. $\frac{3}{16}$

No.123

3" × 3" 14.4 to 33.8 Lbs. $\frac{5}{16}$

No.124

$2\frac{3}{4}$ × $2\frac{3}{4}$ 13.1 to 25 Lbs. $\frac{5}{16}$

No.125

$2\frac{1}{2}$ × $2\frac{1}{2}$ 11.9 to 22.5 Lbs. $\frac{5}{16}$

No.126

$2\frac{1}{4}$ × $2\frac{1}{4}$" 10.8 to 17.8 Lbs. $\frac{5}{16}$

No.127

2" × 2" 7.1 to 13.6 Lbs. $\frac{3}{16}$

No.128

$1\frac{3}{4}$ × $1\frac{3}{4}$" 6.2 to 11.7 Lbs. $\frac{3}{16}$

No.129

$1\frac{1}{2}$ × $1\frac{1}{2}$" 5.3 to 9.8 Lbs. $\frac{3}{16}$

No.130

$1\frac{1}{4}$ × $1\frac{1}{4}$" 3 to 5.8 Lbs. $\frac{3}{16}$

No.131

1" × 1" 2.3 to 4.4 Lbs. $\frac{3}{16}$

No.132

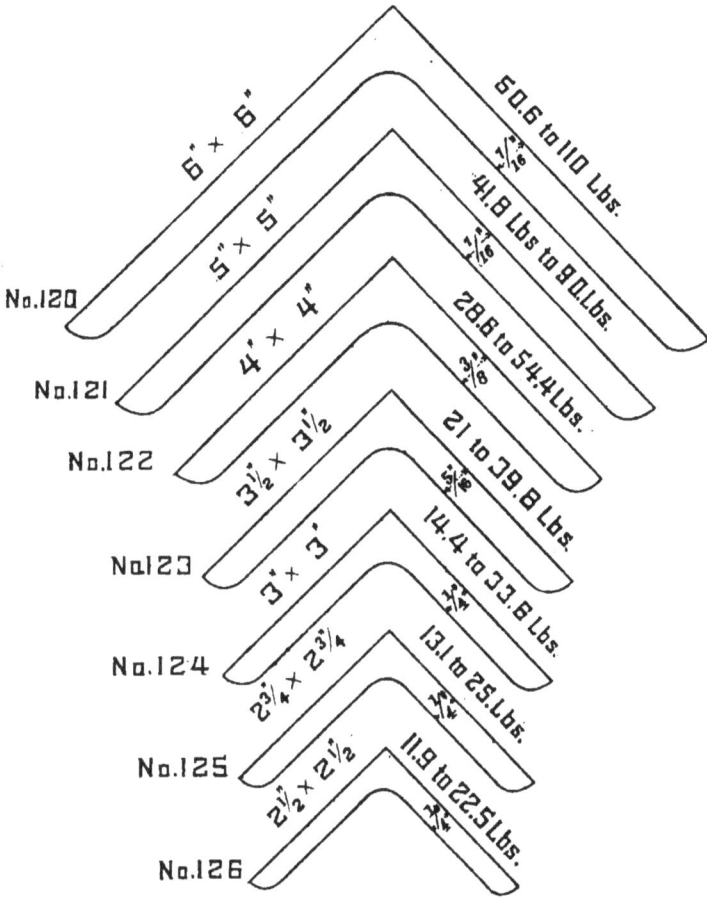

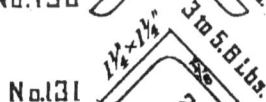

All weights given in pounds per Yard.

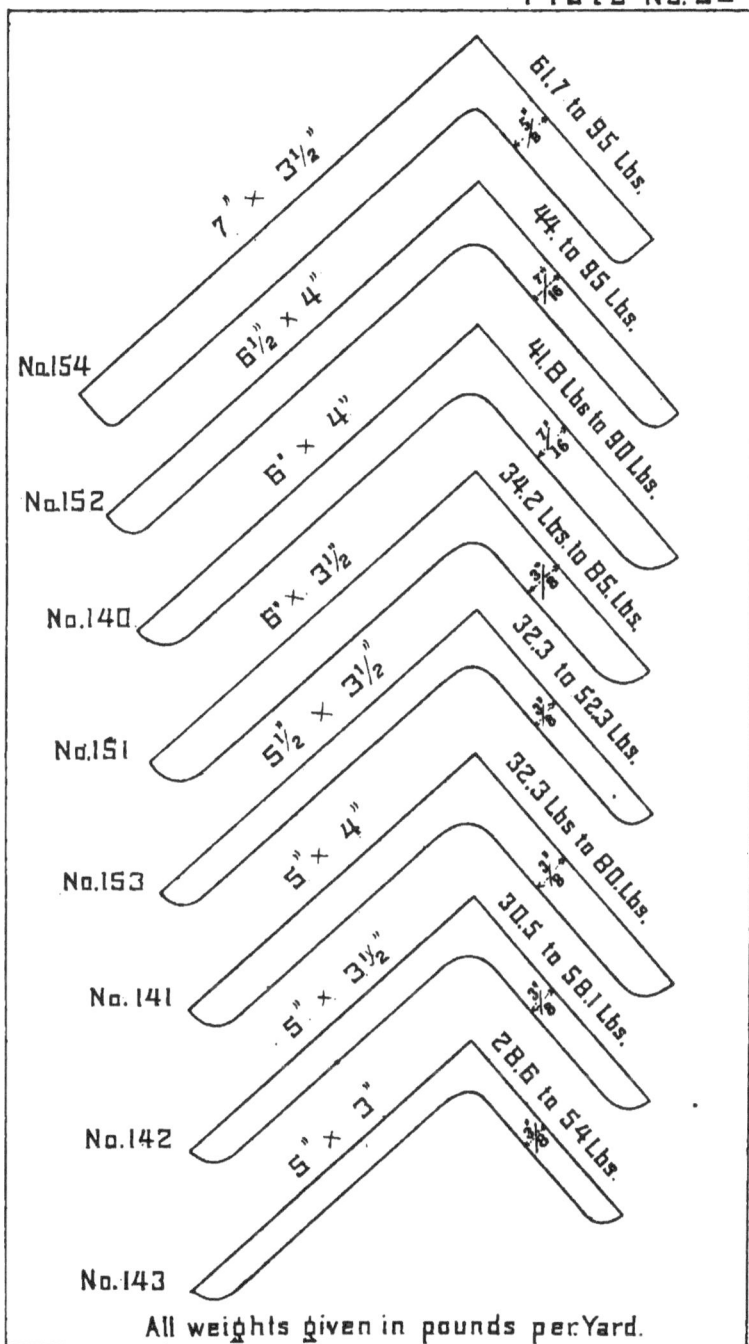

No.154   7" × 3½"   61.7 to 95 Lbs.

No.152   6½" × 4"   44. to 95 Lbs.

No.140   6" × 4"   41.8 Lbs to 90 Lbs.

No.151   6" × 3½"   34.2 Lbs. to 85 Lbs.

No.153   5½" × 3½"   32.3 to 52.3 Lbs.

No.141   5" × 4"   32.3 Lbs to 80 Lbs.

No.142   5" × 3½"   30.5 to 58.1 Lbs.

No.143   5" × 3"   28.6 to 54 Lbs.

All weights given in pounds per Yard.

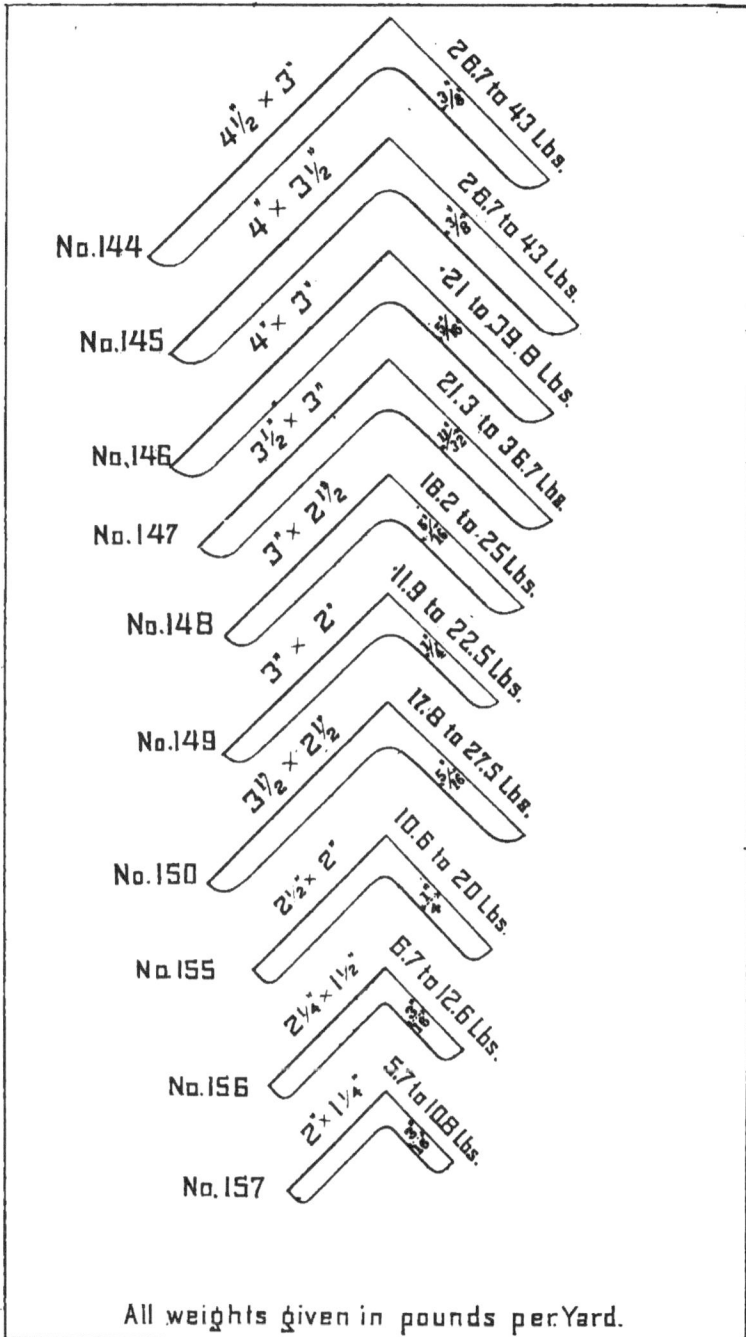

No.144   4½" × 3"   26.7 to 43 Lbs.   ⅜"

No.145   4" × 3½"   26.7 to 43 Lbs.   ⅜"

No.146   4" × 3"   21 to 39.8 Lbs.   ⅜"

No.147   3½" × 3"   21.3 to 36.7 Lbs.   ⅜"

No.148   3" × 2½"   18.2 to 25 Lbs.   ⅜"

No.149   3" × 2"   11.9 to 22.5 Lbs.   ⅜"

No.150   3½" × 2½"   17.8 to 27.5 Lbs.   ⅜"

No.155   2½" × 2"   10.6 to 20 Lbs.

No.156   2¼" × 1½"   6.7 to 12.6 Lbs.

No.157   2" × 1¼"   5.7 to 10.8 Lbs.

All weights given in pounds per Yard.

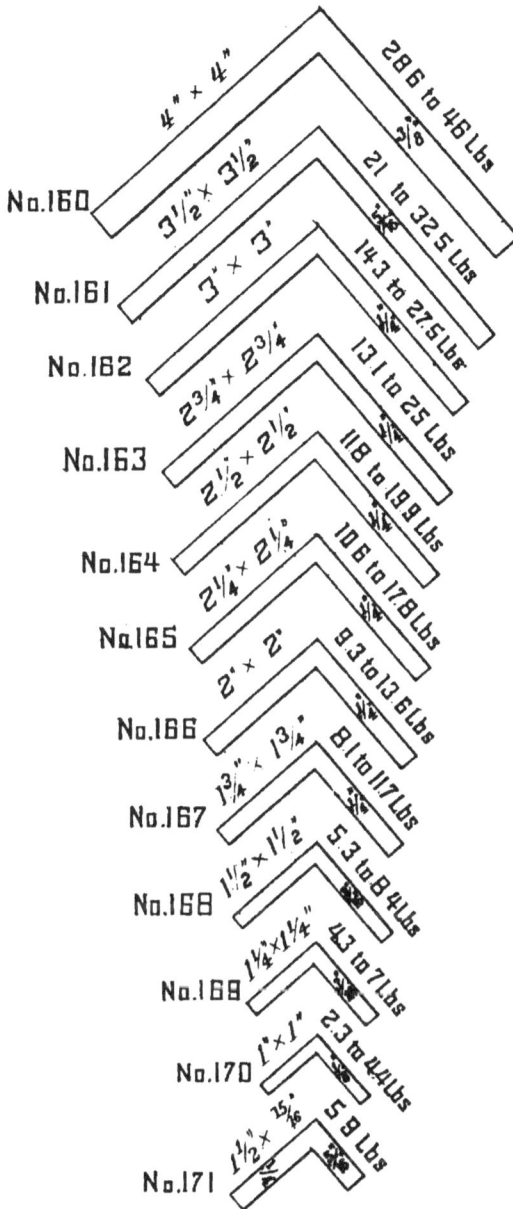

No.160 — 4" × 4" — 28.6 to 46 Lbs

No.161 — 3½" × 3½" — 21 to 32.5 Lbs

No.162 — 3" × 3" — 14.3 to 27.5 Lbs

No.163 — 2¾" × 2¾" — 13.1 to 25 Lbs

No.164 — 2½" × 2½" — 11.8 to 19.9 Lbs

No.165 — 2¼" × 2¼" — 10.6 to 17.8 Lbs

No.166 — 2" × 2" — 9.3 to 13.6 Lbs

No.167 — 1¾" × 1¾" — 8.1 to 11.7 Lbs

No.168 — 1½" × 1½" — 5.3 to 8.4 Lbs

No.169 — 1¼" × 1¼" — 4.3 to 7 Lbs

No.170 — 1" × 1" — 2.3 to 4.4 Lbs

No.171 — 1½" × ¹⁵⁄₁₆" — 5.9 Lbs

All weights given in pounds per Yard.

No. 180   No. 181   No. 182   No. 183   No. 184

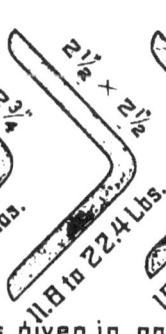

All weights given in pounds per Yard.

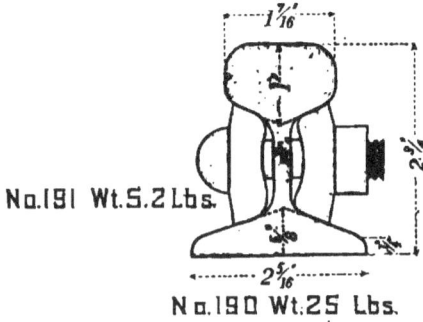

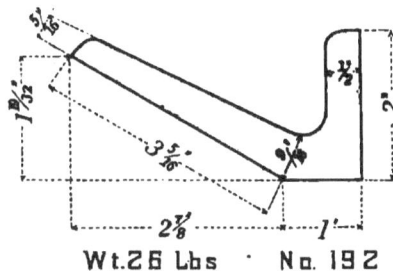

No.191 Wt.5.2 Lbs.

No.190 Wt.25 Lbs.

Wt.26 Lbs · No. 192

No. 194.   No.193.

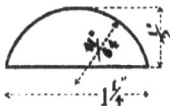

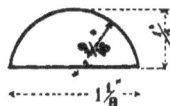

Wt. 4.8 Lbs. p. Yard.   Wt 4.3 Lbs. p. Yard.

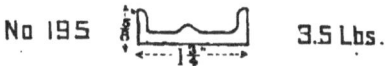

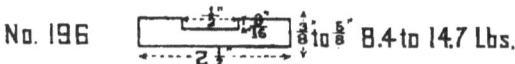

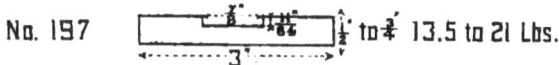

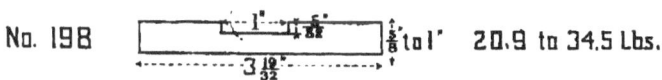

No 195   3.5 Lbs.

No. 196   8.4 to 14.7 Lbs.

No. 197   13.5 to 21 Lbs.

No. 198   20.9 to 34.5 Lbs.

All weights given in pounds per Yard.

METHOD OF INCREASING SECTIONAL AREAS.

Cross-hatched portions represent
the minimum sections, and the
blank portions the added areas.

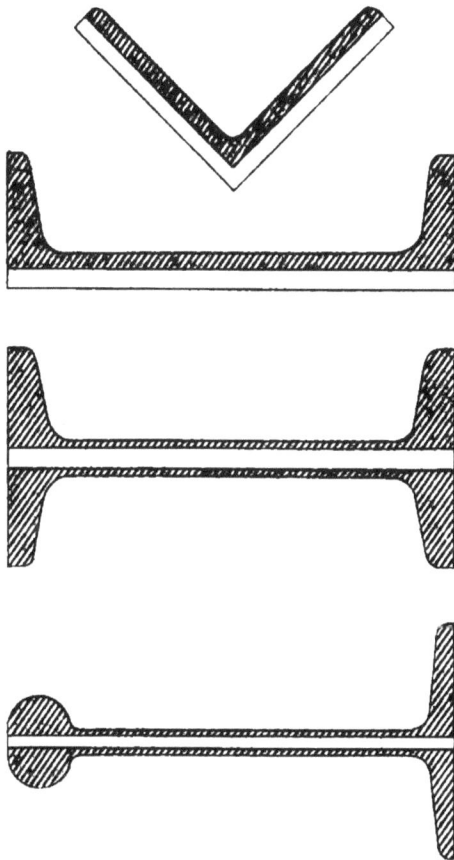

All weights given in pounds per Yard.

www.ingramcontent.com/pod-product-compliance
Lightning Source LLC
Chambersburg PA
CBHW021520210326
41599CB00012B/1330